Mexican Gray Wolves, Courts, and the Border Wall

Mexican Gray Wolves, Courts, and the Border Wall

Lobo Returns from Limbo

Edward A. Fitzgerald

BLOOMSBURY ACADEMIC
NEW YORK • LONDON • OXFORD • NEW DELHI • SYDNEY

BLOOMSBURY ACADEMIC
Bloomsbury Publishing Inc, 1359 Broadway, 12th Floor, New York, NY 10018, USA
Bloomsbury Publishing Plc, 50 Bedford Square, London, WC1B 3DP, UK
Bloomsbury Publishing Ireland, 29 Earlsfort Terrace, Dublin 2, D02 AY28, Ireland

BLOOMSBURY, BLOOMSBURY ACADEMIC and the Diana logo are trademarks of Bloomsbury Publishing Plc

First published in the United States of America 2026

Cover photo illustration: Stephen Rumbaugh.

Library of Congress Cataloging-in-Publication Data
Names: Fitzgerald, Edward A., 1949- author
Title: Mexican gray wolves, courts, and the border wall: lobo returns from limbo / Edward A. Fitzgerald.
Description: New York : Bloomsbury Academic, 2026. | Includes bibliographical references and index.
Identifiers: LCCN 2025034076 (print) | LCCN 2025034077 (ebook) | ISBN 9781666912821 hardback | ISBN 9781666912845 paperback | ISBN 9781978765757 epub | ISBN 9798216267065 pdf | ISBN 9798881890605
Subjects: LCSH: Mexican wolf—Conservation—Law and legislation—Mexican-American Border Region | Mexican wolf—Reintroduction—Mexican-American Border Region | Cattle—Predators of—New Mexico | Livestock—Losses—New Mexico | Border security—Environmental aspects—Mexican-American Border Region
Classification: LCC QL737.C22 F59 2026 (print) | LCC QL737.C22 (ebook) | DDC 599.77309721—dc23/eng/202507128
LC record available at https://lccn.loc.gov/2025034076
LC ebook record available at https://lccn.loc.gov/2025034077

ISBN: HB: 978-1-6669-1282-1
ePDF: 979-8-216-26706-5
eBook: 978-1-9787-6575-7

Typeset by Deanta Global Publishing Services, Chennai, India

For product safety related questions contact productsafety@bloomsbury.com.

To find out more about our authors and books visit www.bloomsbury.com and sign up for our newsletters.

Contents

Abbreviations

AG	Attorney General
APWE	Americans for Preservation of the Western Environment
APA	Administrative Procedures Act
AZGFD	Arizona Game and Fish Department
AZGFC	Arizona Game and Fish Commission
BRWRA	Blue Range Wolf Recovery Area
CBP	Customs and Border Patrol
CBD	Center for Biodiversity
CCC	Catron County Commission
DEIS	Draft Environmental Impact Statement
DHS	Department of Homeland Security
DOD	Department of Defense
ESA	Endangered Species Act
DOW	Defenders of Wildlife
FEIS	Final Environmental Impact Statement
FS	U.S. Forest Service
FWS	U.S. Fish and Wildlife Service
GAO	General Accounting Office
IIRIRA	Illegal Immigration Reform and Immigrant Responsibility Act
INS	Immigration and Naturalization Service

MWEPA	Mexican Wolf Experimental Population Area
NEPA	National Environmental Policy Act
NMGFC	New Mexico Game and Fish Commission
NMGFD	New Mexico Game and Fish Department
NRM	Northern Rocky Mountains
PEER	Public Employees for Environmental Responsibility
PVA	Population Viability Analysis
RI	Rewilding Institute
RIDA	Real ID Act
SBCC	Southern Border Communities Coalition
SEIS	Supplemental Environmental Impact Statement
SFA	Secure Fences Act
SHS	Secretary of Homeland Security
SOD	Secretary of Defense
SOI	Secretary of Interior
SOP	Standard Operating Procedure
WLLDPG	Wolf-Livestock Loss Demonstration Project Grants
WLS	USDA Wildlife Services

Acknowledgments

I want to thank Dr. Ava Chamberlain for her support and her assistance in proofreading and preparing this book's manuscript. I also want to thank Stephen Rumbaugh for his cover design.

Parts of this work have appeared in modified form in several journals. I want to thank the following journals for their permission to reprint portions of these articles.

"New Mexico Cattlegrowers Association v. U.S. Fish and Wildlife Service: Lobo Returns from Limbo," 46 Natural Resources Journal 9 (2006)

"The Lobo Limps on From Limbo: A History, Summary, and Outlook for Mexican Wolf Recovery in the American Southwest," 29 Colorado Natural Resources, Energy & Environmental Review 224 (2018).

"San Diego Border Infrastructure Environmental Litigation: Return of the Walking Dead," 50 Environmental Law 151 (2020).

"Sierra Club v. Trump, California v. Trump: Border Wall Funding Knocked Down," 12 Arizona Journal of Environmental Law and Policy 179 (Spring 2022).

"Lobo's Travails on the Road from Limbo," 53 Texas Environmental Law Journal 1 (Fall 2023).

Introduction

The return of the Mexican gray wolf (lobo), the most endangered wolf subspecies in the world, to the southwest has been controversial. The Mexican wolf was almost exterminated by state bounties and the federal government to serve the interests of the livestock industry, hunters, and developers. State bounties had mixed success, so the federal government engaged in a war against the Mexican wolf. Scientists continually warned about the ecological dangers presented by wolf eradication. The rise of the environmental movement in the 1960s and 1970s resulted in a new perception of the wolf from a wanton killer to a noble ecosystem manager. New statutes were enacted, including the Endangered Species Act (ESA), which protects the Mexican wolf. The new statutes opened up congressional and administrative processes to allow public interest groups to participate in the policymaking process. The new statutes contained citizen suit provisions, which allowed public interest groups to challenge government actions in the courts. Litigation played a major role in the Mexican wolf's reintroduction and recovery.

The Mexican wolf was declared an endangered species in 1976. The ESA required the federal government to recover the Mexican wolf as a functioning component of its ecosystem. Recovery efforts slowly began in the early 1980s. The first recovery plan was completed in 1982 during the Reagan administration. The final rule for the reintroduction of the Mexican wolf into the Blue Range Wolf Recovery Area (BRWRA) in Arizona and New Mexico as a threatened species was published in January 1998 during the Clinton administration. Initially, Mexican wolves were released into the Apache National Forest in Arizona and allowed to disperse into the Gila National Forest in New Mexico. Wolves were not allowed to establish territory beyond the BRWRA. The livestock industry brought suit opposing the reintroduction. After the federal district court upheld the reintroduction, eleven wolves were released into the BRWRA. Lobo has experienced many travails on the road to recovery. The journey has proceeded with various stops and starts and has faced litigation at every step along the way. Judicial decisions have, for the most part, been instrumental in advancing recovery efforts. However, judicial decisions allowing the construction of border barriers and the waiver of environmental laws have posed obstacles to Mexican wolf recovery.

Mexican wolf recovery, like other carnivore conservation efforts, is the tip of the iceberg for many other social and political problems. Scholars point out that the "conservation of large carnivores is also linked inextricably to several broader issues, including the ESA, ecosystem management, use of western public lands." Arguments over carnivore conservation are "often a surrogate for broader cultural conflicts: preservation vs. use of resources, recreation-based economies vs. extraction-dependent economies; urban vs. rural values, and state's rights vs. federal authority."[1]

Lobo's return is important for the southwest ecosystem. The Mexican wolf is a keystone species. Keystone species influence ecosystem functions and have inverse impacts in relation to their numbers. Keystone species trigger trophic cascades, which impact other plants and animals. The Mexican wolf is also a summit predator that sustains biodiversity and maintains ecosystem harmony. The wolf helps manage ungulate prey populations by removing diseased and genetically inferior animals. The wolf prevents the prey population from exceeding the carrying capacity of the area. Overgrazing by ungulates adversely affects the growth of trees and vegetation. It also causes erosion, particularly in riparian areas, which affects water quality. Both overgrazing and erosion adversely affect carbon capture. The wolf provides additional benefits across the ecosystem. After the wolf makes a kill, sustenance is provided for the entire food chain. The wolf maintains an important balance among predators. It limits the number of coyotes, leaving much of the coyote's prey available for other predatory birds and foxes. A balanced ecosystem preserves valuable genes and does not contribute to climate change. In addition, the Mexican wolf requires large interconnected wildlands and wildlife corridors, so it helps to preserve wilderness.[2]

Mexican wolf recovery has faced several major problems. Mexican wolf recovery has generated conflict between economic and environmental interests. It has been opposed by extractive industries, which have dominated western public land policy. Opponents view wolf recovery as a threat to their economic interests. Wolves prey on livestock and possibly threaten ranching on public lands. Wolves take game, which limits hunting opportunities. Wolves occupy wilderness areas, which constrains development. On the other hand, environmental groups support Mexican wolf recovery. Supporters view the

Mexican wolf as the missing component in the southwest ecosystem, which must be restored to maintain nature's balance. In the future, increasing human resource demands and the changing climate will threaten biodiversity and test the ecosystem's resiliency. The recovery and protection of the Mexican wolf, which performs many important ecological functions, is imperative.

Mexican wolf recovery has strained relations between the federal and state governments. Federal wolf recovery efforts pursuant to the ESA preempt state and local laws. The states view this as an intrusion on their traditional authority over wildlife. The states argue that they are more competent to manage wildlife to meet local conditions. In addition, states and localities are deprived of revenues that would be generated by the precluded economic activities.

Mexican wolf recovery raises questions regarding the role of science in political decision-making. Supporters of wolf recovery point out that the ESA requires decisions regarding endangered and threatened species to be made on "the basis of best scientific and commercial data available." Opponents argue that wolf recovery must be integrated with other economic interests.

Wolf recovery involves tribal management. Different tribes have different views regarding wolf recovery. Some tribes support the return of the Mexican wolf as a necessary part of their lost culture. Others oppose recovery because it threatens their livestock and revenues derived from non-tribe hunters on tribal lands.

Wolf recovery is perceived differently by urban and rural residents. Middle-class individuals in urban areas generally support wolf recovery but do not have to deal with the impacts of wolf recovery. Rural residents see wolves as a threat to their economic interests and an unwelcomed federal intrusion into their way of life. In addition, wolf recovery undermines private property, which is the basis of individual freedom.[3]

Mexican wolf recovery is intimately tied to U.S. immigration policy. Its recovery has been hampered by the construction of border barriers built primarily during the George W. Bush and Donald J. Trump administrations. These barriers fragment the habitat, prevent cross-border migration, and complicate cooperative relations with Mexico.

All of these issues have generated public law litigation, which has focused on statutory interpretation, administrative implementation of policy, and constitutional limitations. Various interest groups, both supporters and opponents, have brought suits challenging federal reintroduction and management of the Mexican wolf. Federal courts were called upon to resolve these complex policy disputes through public law litigation, which is both remedy-oriented because it seeks to resolve the conflict between the parties and reform-oriented because it is designed to change administrative and/or congressional policy.[4]

Public law litigation differs from traditional litigation. Traditional litigation is bipolar and focuses on a past event. The remedy sought flows directly from the right violated. The judge is a neutral arbiter applying the law to facts in a one-time event. Public law litigation involves multiple parties. The rules regarding standing and intervention have been broadened to provide interest groups access to the courts. The litigation focuses on present and future government actions. Groups seek injunctions to halt federal action. The judge plays an active role, interpreting statutes and overseeing federal agency compliance with legal mandates. The judge has to consider legislative facts, which are policy oriented, and adjudicative facts, which are party oriented. The litigation is not a one-time endeavor but is often part of the ongoing political controversy.[5]

Supporters of public law litigation, judicial proponents, argue that judicial involvement in public policy disputes is beneficial. It allows groups, whose interests are underrepresented, to challenge administrative action. Public law litigation brings public attention and mobilizes support on both sides of the issue. The neutral judge, who is the legal expert, must decide the issue in the context of a well-reasoned decision. The judge utilizes the traditional tools of statutory interpretation—text, intent, and purposes—to determine the meaning of the law and reviews agency action to assess statutory compliance. The court ensures that the public values manifested in the statutes are realized and executive action comports with congressional will. The court's decision does not necessarily end the process but often serves as a catalyst for subsequent congressional and executive action.[6]

Public law litigation is not universally applauded. Critics, judicial skeptics, argue that the courts should follow the traditional model of litigation and only resolve disputes involving the infringement of individual rights, not conflicts over public policy. They view public law litigation as the extension of interest group politics in the guise of litigation. Groups unsuccessful in the congressional and executive branches utilize litigation to achieve their political goals. Judicial involvement in public policy conflicts detracts from the court's essential function of resolving disputes between individuals, not majoritarian interests. Courts lack the capacity to resolve complex policy disputes. Judges, who are generalists, do not possess the skills and information to comprehend technical and scientific issues and employ the tools of statutory interpretation to substitute their own policy views for those of administrators. The court's episodic involvement interferes with the agency's functions, fragments the process, disrupts the agency's agenda, misallocates scarce agency resources, precludes the necessary political trade-offs, generates costly defensive procedures, and produces bad decisions. From their perspective, federal courts should defer to democratically accountable executive agencies, which possess the requisite policy expertise.[7]

This book examines the reintroduction and recovery of the Mexican wolf to assess the efficiency and impact of public law litigation. Judicial decisions are analyzed to determine whether the federal courts correctly interpreted the statutory and constitutional mandates to advance Mexican wolf recovery.

Chapter 1 provides a theoretical framework, analyzing the role of the courts in the policymaking process.

Chapter 2 examines the history of the Mexican wolf's near-extermination by the state and federal governments in the context of changing federal public land policy.

Chapter 3 examines the livestock industry's effort to stop Mexican wolf reintroduction and recovery. The unsuccessful litigation brought by the livestock industry to stop Mexican wolf reintroduction in New Mexico Cattle Growers Association v. U.S. Fish and Wildlife Service (FWS) and Coalition for Arizona/New Mexico for Stable Economic Growth v. FWS during the Clinton and George W. Bush administrations are analyzed.

Chapter 4 examines the federal government's conflict with New Mexico regarding Mexican wolf recovery from the George W. Bush through the Trump administration. The litigation leading to the 2009 settlement agreement in Defenders of Wildlife (DOW) v. Tuggle that restored the FWS primary management authority; the 2012 litigation in Wild Earth Guardians (WEG) v. Lane upholding the state's action allowing trapping in the BRWRA; and the 2018 litigation in New Mexico Game and Fish Department v. FWS reversing the state policy that prevented the FWS from releasing Mexican wolves into the state are analyzed.

Chapter 5 addresses Mexican wolf management issues, which involved the use of the best available science mandated by the ESA in the Obama administration. The litigation in DOW v. FWS that prompted the administration to change the status of the reintroduced Mexican wolf to endangered subspecies in 2015, the 2015 litigation in Center for Biodiversity (CBD) v. Jewell that halted the revised management regulations for not following the best available science, and the

Biden administration's revisions of the management regulations in 2023 are examined.

Chapter 6 deals with the FWS development of Mexican wolf recovery plans, particularly the use of the best available science. The history of the recovery plans is examined. The litigation in Center for Biological Diversity v. Zinke, Center for Biological Diversity v. Haaland, and WEG v. Haaland extending from the Obama through Biden administrations, challenging the recovery plans, are analyzed.

Chapter 7 examines the construction of border barriers and the secretary of homeland security (SHS) waiver of all laws impeding the construction of border barriers along the US–Mexico border from the George W. Bush through the first Trump administration. The litigation during the Bush administration, challenging the construction of border barriers and utilization of the waiver authority; the litigation regarding President Trump's diversion of funds under the guise of a national emergency to build the border wall; and the litigation In Re Border Infrastructure Environmental Litigation, which upheld the SHS waiver authority, are analyzed. Efforts regarding the border wall during the Biden administration are examined.

The conclusion summarizes the findings and provides recommendations to advance Mexican wolf recovery in the future.

Notes

1 Steven A. Primm and Tim W. Clark, "Making Sense of Policy Process for Carnivore Conservation," 10 Conservation Biology 1036–45 (1996).

2 William J. Ripple et al., "Status and Ecological Effects of World's Largest Carnivores," 343 Science 151 (January 2014).

3 Martin A. Nie, *Beyond Wolves* (University of Minnesota, 2003), 26–67.

4 David M. Trubek, "Environmental Defense I: Introduction to Interest Group Advocacy in Complex Disputes," in Burton A. Weisbrod, Joel F. Handler, and Neil K. Komesar, *Public Interest Law* (University of California, 1978), 153–4.

5 Abram Chayes, "The Role of the Judge in Public Law Litigation," 89 Harvard Law Review 1281 (1976).

6 Owen Fiss, "The Forms of Justice," 93 Harvard Law Review 1 (1979). John Denvir, "Towards a Political Theory of Public Interest Litigation," 54 North Carolina Law Review 1133 (1976).

7 Donald Horowitz, *The Courts and Social Policy* (Brookings Institution, 1977). Michael S. Greve, *The Demise of Environmentalism in American Law* (American Enterprise Institute, 1996). Gerald N. Rosenberg, *The Hollow Hope: Can Courts Bring About Social Change?* (University of Chicago, 1991). Robert A. Kagan, *Adversarial Legalism: The Way of American Law* (Harvard University, 2001).

1

Theoretical Framework

This chapter discusses the underlying theoretical issues in the Mexican wolf litigation. Mexican wolf reintroduction and recovery involved the interaction of the three branches of the federal government within the framework of the Constitution. Public law litigation raises interesting questions regarding the role of each branch: Has Congress sufficiently articulated the statutory mandate? Should the courts or administrative agencies be principally responsible for defining the legal mandates? What tools should be used by the court to define the statutory mandate? Can the president assume authority over congressional appropriations? Is the grant of unlimited waiver authority consistent with the Constitution? The answer to these questions depends to a degree on the court's assessment of the congressional and executive action. The debates between judicial skeptics, who call for judicial restraint regarding executive decisions, and judicial proponents, who support a more active judicial role in overseeing executive actions, on these issues are presented. Finally, the impact of public law litigation on the ongoing institutional dialogue regarding public policy is discussed.

1.1 Constitutional Setting

Policymaking can be defined as "a course of action or inaction rather than specific decisions" that proceeds through a series of interrelated processes: agenda setting, policy formulation, implementation, and assessment. Policymaking occurs within the constitutional framework, which separates power between the legislative, executive, and judicial branches; establishes a system of checks and balances; and divides power between the federal and state governments. Congress makes the law, the Executive implements the law, and the judiciary adjudicates disputes. The vertical separation of powers between the three branches promotes efficiency and prevents the arbitrary exercise of power by any branch. The horizontal separation of power between the federal and state governments prevents tyranny and preserves liberty.[1]

The constitutional structure requires governmental institutions to cooperate and collaborate with each other. Governmental institutions also compete with each other, attempting to read the signals sent by other institutions to maximize their opportunity for success. The policymaking process is "a spiral of institutional competition and cooperation," which provides numerous access points for competing interest groups to attempt to influence policy.[2]

Judicial skeptics argue that the Constitution provides for stability and continuity in governance. The separation of powers restrains the federal government from interfering with the authority left to other spheres. The people are represented by elected officials, not unelected judges. Elected officials are accountable for their decisions at the ballot box, not in federal courts.[3]

Judicial proponents counter that there are defects in the legislative and administrative process that exclude the consideration and implementation of important public values. There are differences between the law in action and the law on the books. Public law litigation addresses these problems and gives meaning to important public values inherent in the constitutional structure.[4]

1.2 Congress

Congress plays a key role in public policy. Congress is a bicameral institution whose members are elected by distinct geographical constituencies. Congress helps to set the agenda by calling attention to problems through hearings, the introduction of bills, and speeches. Congress formulates policies and enacts them into law. The legislative process is decentralized and very fragmented. The majority party in each house, which controls the leadership positions and every committee, plays a crucial role. Legislation is drafted and altered in the committees, on the floor of each house, and revised by a conference committee. Legislation must be approved by both houses and signed by the president, who is often actively involved in the process. There is logrolling, political posturing, manipulation of procedures, and strategic voting throughout the process. Legislative decisions are the result of competing political pressures and are driven by two-, four-, and six-year election cycles.

There are contending views of the legislative process. Public choice theory considers the legislature to be a marketplace in which public goods are distributed to private interests. Legislation is an economic transaction. Legislators, who are seeking reelection, enact policies that benefit private interests in return for votes, campaign contributions, and political influence. Interest groups compete for legislation, which will distribute public benefits to their members. The greater the benefits or costs, the more time and money groups will spend to achieve their goal. There is little demand or incentive to enact public interest legislation with dispersed benefits or costs. Such legislation is plagued by "free riders," who will not participate or invest in its enactment. Legislation does not further "some general social goal," but embodies deals that reflect private preferences. There is no deliberative process striving toward the achievement of the public interest.[5]

Civic republicanism posits that Congress is a deliberative body in pursuit of the "social good." Interest groups respond to problems and bring diverse perspectives to the deliberative legislative process, which is characterized by dialogue and discussion. Congress is the "forum for identifying or defining [objectives], and acting towards those ends." The legislative process is "one of mutual search through joint deliberation" in which "moral insight, sociological understanding, and goodwill are legislative virtues." The end product is legislation that realizes the public interest.[6]

The deliberative process is reinforced by proper procedures. Henry Hart and Louis Sacks point out that "the substance of a decision cannot be planned in advance in the form of rules and standards," but "the procedure of decision commonly can be." Proper procedures ensure that all views are heard; factual information is subject to critical scrutiny; and the benefits and costs are adequately evaluated. Hart and Sacks assert that "the best criterion for sound legislation is the test of whether it is the product of a sound process of enactment."[7]

1.3 Executive

The president influences the political agenda by raising issues, structuring the public debate, and endorsing or discouraging policies. The president shapes the formulation of policy by generating proposals, mobilizing support for their enactment, and signing or vetoing legislation. The president's success is often determined by the partisan alignment in Congress, his standing in the polls, administrative skills, and leadership ability.[8]

Executive agencies, which are established by Congress, play a role in agenda setting and policy formulation by providing information that serves as the basis for executive and legislative actions. Executive agencies implement statutory policies. Congress generally delegates broad administrative authority to the agency because of election concerns, the need to achieve legislative compromises, the complexity of the problems, and unknown future contingencies. The agency implements statutory policies through rules and regulations enacted pursuant to the enabling statute or the Administrative Procedures Act (APA). Agency action proceeds through both formal and informal rulemaking. Formal rulemaking involves an adjudicatory hearing, followed by a decision based on substantial evidence. Informal rulemaking, also known as "notice and comment," involves the publication of the proposed agency action in the Federal Register. Public comment is taken on the proposal. The final agency decision, which addresses the public comments, is published in the Federal Register. Policymaking through informal rulemaking has been described as "one of the greatest inventions of modern government."[9]

Executive agencies are supplementary lawmakers under conflicting pressures. Who controls agency action is at the center of the debate regarding public law litigation. The agency responds to the president, who nominates agency officials and sets the direction of national policy. The Office of Management and Budget (OMB) reviews proposals for legislation, controls the agency's initial budget proposal, monitors proposed rules for compliance with Executive Orders, and communicates the White House position on important matters.[10]

Judicial skeptics argue that the president, who is democratically accountable, should police agency actions. The president represents a national constituency, so is not constrained by geographical interests. The president can enforce priorities and provide cohesion. Judicial proponents counter that presidential oversight of administrative action is limited. The president has many responsibilities, so is not involved in most administrative decisions. Presidential appointments are usually done to service interest groups and/or congresspersons. Once appointed, administrators are constrained by statutory mandates and their constituencies. The president primarily influences agency action by providing a sense of direction.[11]

Executive agencies are subject to continued congressional influence. Congress defines the agency's jurisdiction and procedures, confirms its leadership, and approves its budget. Specialized committees oversee the agency's statutory implementation and review the agency's budget. Budgetary politics is a common means for formulating and influencing the implementation of policy. Congressional action continues, changes, or ends the policy.[12]

Executive agencies are driven by their own internal dynamics. The agency serves the groups affected by its decisions. Interest groups provide information and generate political support for the agency with legislators. A symbiotic relationship develops between the agency and its clientele, which, unlike the public, are well-organized and maintain constant contact with the agency. Executive agencies are motivated by the bureaucratic desire to maximize their institutional importance and budget and minimize any threat to their functions.[13]

1.4 Federal Courts

Federal courts affect policymaking through "the establishment and application of authoritative rules," which influence government actions. Contending interest groups unsuccessful in pursuing their policy demands in the legislative and executive arenas continue their struggle in the federal courts through public law litigation. The courts are called upon to evaluate the executive agency's implementation of policy. The federal courts play an important role in interpreting the statutory mandates and reviewing executive actions to ensure they conform to constitutional restrictions and statutory mandates and are not arbitrary and capricious. In addition, public law litigation advances other group interests, such as focusing public attention on the issue, enlisting the support of other political actors, and activating other groups.[14]

Federal courts are hierarchically organized. There are ninety-four federal district courts, thirteen appellate courts, and the Supreme Court. Federal district courts serve primarily as dispute resolvers. Federal appellate courts supervise the lower courts' decisions and promote uniformity within the circuit. Appellate courts are more concerned with the development and enunciation of broad rules and policies, rather than merely resolving the disputes between litigants. Appellate decisions are generally the final word on the issue because the Supreme Court only reviews a limited number of cases.[15]

1.4.1 Statutory Interpretation

One of the court's major functions in public law litigation is to interpret statutes. Statutory interpretation is dependent on the court's view of the legislative process. There are different views regarding the relationship between the courts and Congress. Public choice adherents contend that courts are part of interest group politics. Interest groups invest political capital and make legislative deals that favor their members. Groups are willing to invest time and capital for a long-term deal, which is essentially a contract between themselves and legislators. Groups are aware that future administrations and Congresses will not be committed to the past deal, so they need a way to ensure that the deal will be enforced in the future. Judicial review by an independent court provides the means to ensure the longevity of the legislative deal. There is no place for public interest, which is anathema to interest group politics.[16]

Civic republicans take issue with this view. Civic republicans argue that the Constitution bars the naked transfer of public goods to interest groups through politics. The court's role is to discourage interest group deals by raising their costs. The court must look past the deal and promote the public interest in the statute. Judicial review makes Congress accountable to the voters by making legislation transparent, exposing hidden deals, and enforcing only the provisions consistent with statutory purposes and policies.[17]

There are different theories of statutory interpretation, which are the subject of an ongoing debate regarding democratic legitimacy.[18] This study adopts the traditional position, originalism, which recognizes a statute as the command of the sovereign that must be interpreted by other agencies of government. Originalism presumes that the democratically elected legislature has the constitutional responsibility for making policy and that the legislative command is supreme. The court, acting as the agent of the legislature, determines the legislative mandate by examining the text, intent, and purpose of the statute. The court, which is not accountable to the electorate, has some discretion but is constrained to act within the parameters of the statute to carry out its purposes. Originalism is a positivist approach that is consistent with the constitutional principles of popular sovereignty, majoritarianism, separation of powers, checks and balances, and federalism.[19]

An originalist interpretation begins by examining the text of the statute, which has been enacted into law through the constitutionally prescribed process. The text is the best evidence of legislative intent. Reliance on the text confines the court's inquiry; increases the probability of obtaining judicial agreement in particular cases; and provides certainty and predictability in the law.[20]

There are different textual positions. Strict textualism does not envision the court as an agent of the legislature, but of the Constitution. Statutes are enacted through constitutional process. Only words are enacted into law. Strict textualism rejects legislative intent and purposes that only serve as vehicles for judges to read their policy preferences into the statute. Justice Scalia noted that "judges interpret laws rather than reconstruct legislators' intentions. Where the language of those laws is clear, we are not free to replace it with an unenacted legislative intent."[21] Strict textualism incorporates the skepticism of public choice theory and mandates that statutes be strictly interpreted to decrease the public benefits dispensed to private groups by legislators seeking reelection.[22]

Strict textualism poses too many interpretative problems. First, the text alone is generally insufficient to determine statutory meaning because it is often ambiguous, vague, or incomplete. Because the words do not have clear meaning outside of the context in which they are used, external sources must be utilized to determine the meaning of the statutory language.[23]

Second, strict textualism does not foster a sensible reading of the statute that is designed to carry out the legislative purpose. Instead, particular language is isolated from the legislative enterprise of which it is a part. Congress is unable to send messages to the court regarding agency behavior. This deprives the court of valuable information necessary to interpret the statute and frustrates the court's ability to carry out the design of the enacting legislative coalition. Strict textualism increases agency discretion and favors the executive branch in its competitive struggle with Congress over the formulation and implementation of public policy. Strict textualism is a political strategy for disciplining legislatures and judges.[24]

Third, strict textualism is built on the "benign fiction" that rational legislators know the canons of statutory construction, the prior judicial precedents, the existing statutory terrain, and write clear text. There is no support for these assumptions. Congressional staff and interest groups draft the text as well as the legislative history. Judges can manipulate the statutory text as easily as the legislative history.[25] Ignoring legislative history allows the court to assert sole dominance in the interpretative arena.

Strict textualism views the text as the only legitimate interpretive source. Strict textualism attributes meaning to words solely within their statutory terrain by examining dictionary definitions, grammatical constructions, and the canons of statutory construction. The structure of the statute is analyzed to see how the words are utilized and how the definition fits in with the statute as a whole. The relationship of the statute with other statutes and the use of the words in other statutes are also relevant. Weak textualism considers the statutory language to be the best guide regarding legislative intent or purpose.[26]

Since the text generally does not answer the interpretative problem, originalism requires the court to search through the "ashcans of the legislative process" to discover how the enacting legislature would have resolved the question. This inquiry can be either archaeological or hypothetical. An archaeological search entails a review of the text and the legislative history to determine the specific legislative position regarding the issue. A hypothetical inquiry is a reasonable projection about what the original legislature would have said if it possessed the foresight to anticipate what has happened since enactment.[27]

The court must be careful to distinguish between legitimate and manufactured legislative history. Legitimate history is comprised of statements that are designed to explain the meaning of the statute and build support for its enactment. These statements "illuminate the consensus view of the legislature where that view is left in doubt by the enacted language." Manufactured history consists of statements that are not designed to generate support for the statute, but appeal to particular constituencies and influence the judicial interpretation of the statute. These statements do not reflect the consensus of the legislature.[28]

Judicial skeptics are particularly wary of the use of legislative history, which enhances the power of lobbyists, committees and subcommittees, and staff. Legislative history elevates the views of committees and individuals over those of the Congress as a whole and enhances the power of committees over executive agencies. This is particularly true when Congress and the Executive are controlled by different political parties. Justice Scalia argued that the use of legislative history is like looking over a crowd and picking out your friends.[29]

Judicial proponents counter that legislative history reflects the way Congress does business. Congress performs its responsibilities through committees, whose staff writes reports explaining what has been done. There is little danger that information slips past the full Congress. Judge Wald stated, "legislative history is the authoritative product of the institutional work of the Congress. It records the manner in which Congress enacts its legislation, and it represents the way Congress communicates with the country at large." Ignoring legislative history criticizes "the way a coordinate branch conducts its operations and, in that sense, runs the risk of violating the spirit if not the letter of the separation of powers."[30] Justice Breyer declared that "using legislative history to help interpret unclear statutory language seems natural. Legislative history helps a court understand the context and purposes of a statute."[31]

William Eskridge provides a hierarchy of legislative sources that is based on their comparative reliability. First, the most reliable sources are the committee reports, which represent the "collective understanding of those Congressmen involved in drafting and studying proposed legislation." Congress acts through committees and "committee reports should be looked to for the most coherent, thorough, and authoritative explanation of a bill's purpose and intended meaning." The reports show areas of agreement between the House and Senate and issues of disagreement that are resolved by the conference committee.

Second, the "statements by sponsors and/or floor managers," who know the language, intent, and purpose of the statute, are important because other congresspersons defer to their judgment. Third, rejected proposals are relevant as "direct evidence that Congress considered an issue and agreed not to adopt a specified policy." Fourth, the statements of legislators in committee hearings and on the floor, while not as authoritative as the sponsors, show "the general assumptions made at the time a law was enacted." Fifth, statements by nonlegislative drafters and sponsors are generally not important, unless "the statute was a careful compromise" or "there is virtually no other evidence." Finally, legislative silence and subsequent history are accorded little weight, unless "the precise intent of the enacting Congress is obscure."[32]

Judicial skeptics assert that there can be no legislative intent because there is no lawmaker. The preferences of individual legislators cannot be aggregated into any collective intent because their psychological and political motivations cannot be known. Giving effect to legislative intent violates the constitutional requirements of bicameralism and the Presentment Clause that are necessary for the enactment of law. Even if legislative intent exists, it does not bind the court because the legislature's function is to pass statutes, not to apply the law in a particular factual situation. Finally, legislative intent is irrelevant because the statutory language limits the court.[33]

Reed Dickerson rebuts this criticism and points out that all statutes rest on legislative intent, even though it might be difficult to discern. The best approximation of legislative intent is the statute, read in its proper context, supplemented by other reliable extrinsic evidence. Respecting legislative intent places the court in the proper deferential framework regarding the legislature and establishes a criterion of reliability that helps the court select and weigh elements of the language and legislative context.[34] Justice Breyer concurred, arguing that legislative history is a judicial tool employed to resolve complex problems of judicial interpretation. It helps the court to make sense of the statute and allows the court to produce a coherent and consistent body of statutory law.[35]

Public choice advocates also find legislative intent important. The court's role is to carry out the deals negotiated by the legislative coalition. This encourages future negotiations and leads to "stable and predictable methods of interpretation."[36]

An originalist interpretation is also guided by the legislative purpose, which is the ultimate motive of the legislature. Henry Hart and Louis Sacks assert that "every statute must be conclusively presumed to be a purposive act." The legislative purpose is the best justification that can be attributed to the statute in terms of its relationship with the set of legal norms operating at the time of the court's decision. The legislative purpose helps the court to determine the legislative intent, directs the court when intent has not been manifested, and allows the court to keep the statute in harmony with contemporary values.[37]

Judicial skeptics argue that reliance on the legislative purpose is indeterminate for several reasons: First, since the legislative purposes are stated in general terms, the court has too much discretion. Second, a statute reflects many different purposes because it must be acceptable to a wide range of interests to be enacted. The different purposes reflected in the political bargain cannot be aggregated into any particular public purpose. Third, focusing on the legislative purpose ignores the statutory text, which is the most legitimate source of legislative intent and is relied upon by the citizens. Finally, the optimistic assumptions regarding the purposeful legislative process are naive.[38]

Hart and Sacks address their critics, asserting that the attribution of purpose does not grant the court unbridled discretion. The court is constrained by the words of the statute, which are enacted by the legislature. The court must be careful not to give the words "a meaning that they will not bear." Words are not empty vessels into which the court pours meaning but have a dual role "as guides to the attribution of general purpose and as factors limiting the particular meanings that can properly be attributed." The words must not be given "a meaning which would violate any established policy of clear statement." The court must "try to put itself in imagination in the position of the legislature which enacted the measure" and assume that the "legislature was made up of reasonable persons pursuing reasonable purposes reasonably." The court asks, "why would reasonable men, confronted with the law as it was, have enacted this new law to replace it?" The court looks for the "mischief" in the prior statute and "the true reason of the remedy" provided by the new statute. Legislative history must be examined "for the light it throws on the general purpose."[39]

Justice Breyer noted that a purpose-oriented approach is consistent with democratic values. It allows the court to better comprehend and realize the legislative goals in the statute, which reflect the public will. Judges should put themselves in the shoes of the theoretical "rational legislator" and imagine how they may have answered the question presented.[40]

Public choice advocates find merit in Hart and Sacks' approach. Private interest legislation is generally based on a public purpose. An interpretative strategy that relies on public purpose ensures this purpose is carried out. The court places "subtle pressures on Congress to act in ways that benefit the pubic" and forces the Congress "to spell out more clearly the terms of its bargains with special interest groups." This raises the cost of interest group activity. Scholars note that "Hart and Sacks were not so much Panglossian empiricists as savvy normativists—crafters of assumptions that provide useful judicial and administrative side-constraints upon the less attractive features of politics."[41]

1.4.2 Judicial Review of Administrative Decisions

Another important role for the courts in public law litigation is to oversee the agency's legal and policy decisions to ensure compliance with the statutory mandates. There is an ongoing debate regarding the interpretive authority of the courts and administrative agencies.

The different positions regarding judicial-agency relationships can be aggregated into the deferential and supervisory models of administrative law. Judicial skeptics champion the deferential model, asserting that courts should defer to agency decision-making for several constitutional reasons. First, the Executive is mandated to "take care that the laws be faithfully executed." Second, Congress delegates its legislative authority to executive agencies to implement the statute. Third, federal agencies when implementing the statute engage in a political balancing process that belongs to the democratically accountable executive branch.[42]

Judicial skeptics argue that the deferential model is based on institutional competence. Executive agencies are more competent than courts to interpret statutes. Agencies have superior knowledge about legislative intent, are familiar with the problems being addressed, participate in the legislative process, know the consequences of the legislative action, and have better interpretive tools (i.e., staff, analytical skills, political sophistication). Agency decisions, which are based on technological and policy concerns, are superior to judicial decisions, which are constrained by principles. Agency decisions provide for geographic uniformity, continuity, harmony, and integration of policies, and quicker and more efficient implementation of the law.[43]

Skeptics assert that adjudication is piecemeal. Courts are reactive and must respond to the issues presented by the parties. Courts are not competent to resolve complex policy disputes that go beyond the declaration of rights and involve the implementation of rights through the oversight of the government bureaucracy. Judges are generalists, whose recruitment, socialization, and role precludes the development of the specialized knowledge involved in complex policy disputes. Courts are well suited to deal with party-oriented adjudicatory facts, but not with policy-oriented legislative facts. The necessity to provide a remedy for the violation of the rights infringed limits the court's ability to design and implement remedies to resolve complex policy disputes. Furthermore, courts do not provide for an ongoing review of policy.[44]

Skeptics point out that judicial involvement in complex policy disputes is based on inadequate knowledge of public bureaucracies. Courts do not understand economic and political constraints on agencies, the implications of various policy options, and the necessity for trade-offs. Policymaking is done under conditions of uncertainty. Scientific and technological decisions involve much speculation. Judicial involvement interferes with policy implementation and "ossifies" the federal bureaucracy. Judicial decisions deflect policy concerns into narrow areas, paralyze policy innovation, foster contradictory policies, misallocate scarce resources, and force agencies to adopt policies that are often inimical to the public interest.[45] Martin Shapiro argues that "policy making via litigation means that the judge enters highly complex ongoing situations just far enough and just long enough to pull out a crucial brick and then scampers away, leaving it to the rest of the political system to somehow make sure that the wall does not fall down."[46]

Judicial skeptics are critical of liberal standing rules, which allow interest groups to utilize the courts to challenge political decisions. To establish standing, the plaintiffs must show that they will suffer an injury in fact, that the injury is caused by the defendant, and that the court is capable of providing redress for the injury. The plaintiffs must also demonstrate that the injury falls within the zone of interests to be protected by the statute.[47] Skeptics complain that the expansion of standing has made courts into just another forum for interest group politics.[48] Only a litigant who suffers an injury to her common law interest should have standing to challenge government action. Individual harm is axiomatic to litigation. Public law litigation is parasitic on the courts' function as a dispute resolver and norm enforcer. Granting interest groups standing to challenge government policy interferes with the Executive's implementation of the law. Statutory beneficiaries should not be allowed to bring their complaints regarding policy implementation to the courts. Instead, they should focus their efforts on the congressional and executive branches.[49]

Judicial proponents argue that the deferential model fails to recognize that the implementation of statutory mandates is a political process. Agencies make choices when interpreting and implementing statutes. Agencies are not disinterested bureaucrats serving the public interest, but political actors under conflicting pressures, which often preclude statutory compliance. Courts oversee agency decisions "to see that important legislative purposes, heralded in the halls of Congress, are not lost or misdirected in the vast hallways of the federal bureaucracy."[50]

Proponents support the supervisory model, which recognizes the important role of the court in overseeing administrative actions and advances the constitutional separation of powers and checks and balances. Judicial review ensures that the executive branch faithfully executes the law and acts within the parameters of the statute to carry out its purposes. Judicial review prevents agency capture and keeps the agency accountable to Congress and ultimately to the voters.[51]

Proponents point out that the supervisory model is supported by the APA, which declares that "the reviewing court shall decide all relevant questions of law, interpret constitutional and statutory provisions, and determine the meaning or applicability of the terms of an agency action." The court is also required to review the agency's factual and policy decisions to ensure that the agency has not acted in a manner that is "arbitrary and capricious, an abuse of discretion, or otherwise not in accordance with law." Furthermore, the APA grants parties aggrieved by federal government action generic standing to challenge federal administrative action.[52]

Proponents contend that judicial review promotes the legitimacy of administrative actions, which is defined as the "ensurance of legality, protection against arbitrariness and selectivity, promotion of procedural regularity, and ensurance against the twin evils of factional tyranny and self-interested representation."[53] Courts maintain the integrity of the system. Louis Jaffe stated, "the availability of judicial review is the necessary condition, psychologically if not logically, of a system of administrative power which purports to be legitimate, or legally valid." Courts serve as "a constant reminder to the administration and a constant source of assurance and security to the citizen."[54]

Proponents argue that the supervisory model recognizes the court's institutional advantages. The court must respond to the complaint. The judge, who is insulated from narrow political pressures, operates under "a professional ideal of reflective and dispassionate analysis of the problem before her and is likely to have had some experience in putting ideal into practice." The judiciary, which is not bureaucratic, can tap energies and resources in and outside of the government to seek a solution that can be tailored to meet the needs of the particular situation. National policy can be applied in limited situations. Litigation permits a high degree of participation by concerned parties without providing them with a veto and generates sufficient information to ensure the proper critical review. Judicial review is an effective mechanism for registering and responding to grievances generated by the operation of public programs in the regulatory state.[55]

Judicial proponents applaud the expansion of standing rules, which allow interest group access to courts. The citizen suit provision in the ESA allows any person to bring suit to stop any person, including the federal government, from violating the statute.[56] General citizen standing is provided in the APA. The Supreme Court has recognized that harm to "scenery, natural and historic objects and wildlife . . . may amount to an 'injury in fact' sufficient to lay the basis for standing under the APA."[57] The Supreme Court also acknowledged that an attenuated line of causation between the alleged environmental harm and the federal action can also support interest group standing.[58] These changes promoted the growth of public law litigation.

Judicial proponents argue that statutory beneficiaries can be harmed by the agency's implementation of the law. Standing can't be limited to those suffering an injury to their traditional common law interests. Congress created new statutory benefits that are entitled to protection. The agency is sometimes reluctant to implement the congressional mandate. Congress granted statutory beneficiaries standing to ensure that the agency follows the law. The citizen suit is a corrective action. The court is not forcing the agency to carry out policies not enacted in the political process, but ensures that the executive branch complies with statutory mandates.[59] Even some judicial skeptics recognize that "standing to obtain judicial review is a critical determination of a party's ability to participate effectively in agency decision-making process." Citizen suits require administrators to "respond to arguments made by parties that can challenge policy decisions in court . . . but they can ignore with relative impunity arguments made by parties that lack this power."[60]

1.4.2.1 Judicial Review of Questions of Law

The courts are sometimes asked to review the agency's interpretation of the law in public law litigation. The courts have the constitutional authority to say what the law means. Alexander Hamilton in Federalist No. 78 stated "the interpretation of the laws is the proper and peculiar province of the courts." Judges determine "the meaning of any particular act proceeding from the legislative body." Justice Marshall reiterated this view in Marbury v. Madison, commenting that "it is emphatically the province and duty of the judicial department to say what the law is."[61] In the past, the courts adopted a case-by-case approach regarding the deference accorded to an agency's legal determination, considering such variables as "an agency's statutory mandate, its familiarity with the regulated industries, its expertise concerning the subject matter, and the type of interpretative problem at issue."[62]

This ad hoc approach was questioned in Chevron v. Natural Resources Defense Council (Chevron). The Supreme Court developed a two-step process regarding judicial review of an agency's legal interpretations. The court must first examine the text and legislative history of the statute to determine "whether Congress has directly spoken to the precise question at issue." If Congress did not directly address the question, the court could "not simply impose its own construction on the statute." The court must move to the second step to determine "whether the agency's answer is based on a permissible construction of the statute." The court did not have to conclude that the agency's construction was "the only one it permissibly could have adopted to uphold the construction, or even the reading the court would have reached if the question had arisen in a judicial proceeding." Instead, the court must defer to "a reasonable interpretation made by the administrator of the agency."[63]

Judges were admonished not to read their policy preferences into the law because they lacked expertise and were not accountable to the electorate. Congress delegated its power to the agency, which could "properly rely upon the incumbent administration's views of wise policy to inform its judgments." The president then makes policy choices that "resolve the competing interest which Congress itself either inadvertently did not resolve, or intentionally left to be resolved by the agency charged with the administration of the statute in light of everyday realities."[64]

Judicial skeptics applauded Chevron because it prevented the federal courts from imposing their views on the agencies. Chevron granted the president more authority and discretion to address problems. Agencies had more flexibility to change their statutory interpretation "in light of new scientific, industrial, or other developments or even because a recently elected administration has a new regulatory program." The development of national policy was facilitated by precluding different circuit courts from rendering different decisions on the same issue, preventing the "balkanization of federal law."[65]

Skeptics argued that agencies would be "encouraged to take more responsibility for interpreting the statutes they implement." Litigants would "make their best arguments, clearly and aggressively, before the agency rather than waiting for the main event in the courthouse."[66] Congress would be encouraged to write clear statutes.

Judicial proponents were critical of Chevron, which they argued violated the separation of powers by granting the agency unbridled discretion to enact executive policy. There was no judicial check on administrative agencies unless a clear signal had been sent by Congress, which never acquiesced to this transfer of authority.

Proponents asserted that agencies lacked the institutional capabilities to determine legal questions. Chevron failed to distinguish between the delegation of interpretative powers and statutory ambiguities. Chevron was based on a misconception of statutory interpretation. The question was not whether Congress had precisely addressed the issue, but whether the statute required or forbade administrative action. Chevron failed to recognize that interest groups would continue to dominate executive agencies. Judicial review was necessary to prevent self-interested decisions: "foxes should not guard henhouses."[67]

Proponents contended that Chevron was inconsistent with evolution of the nondelegation rule, which posits that Congress cannot delegate legislative authority to agencies without providing adequate standards for its execution. The nondelegation rule reinforces the separation of powers and prevents the concentration of power in the Executive. The Supreme Court abandoned the rigid nondelegation rule with the assurance that federal courts would watch over executive power. Executive authority must be constrained by adequate administrative procedures and judicial review. Congress delegated power to executive agencies to make policy within statutory limits, but not to define those statutory limits.[68]

Proponents argued that Chevron was overbroad. Each case presents different circumstances with different degrees of complexity. Justice Breyer pointed out that Chevron asks judges to develop a psychologically troubling perspective. After examining the legal question in depth, it is difficult for the judge to believe both that the agency interpretation is legally wrong and its interpretation is reasonable. Justice Breyer maintained that "the statute's text, its context, the structure of the statutory scheme, and canons of textual construction" as well as "statutory purposes, including those revealed in part by legislative and regulatory history" are relevant in determining whether the statute is ambiguous.[69]

Chevron did not repudiate the supervisory model because the courts employed Chevron in a strong or weak manner. A strong reading required the court to defer to the agency's legal interpretation unless Congress specifically addressed the legal issue. A weak reading stressed the continued use of the traditional tools of statutory interpretation. Only when these tools failed to yield an answer to the legal question was the court required to defer to the agency's legal interpretation.[70] The Supreme Court in Chevron specifically declared that "the judiciary is the final authority on issues of statutory construction and must reject administrative constructions which are contrary to clear congressional intent. If a court, employing the traditional tools of statutory construction, ascertains that Congress had an intention on the precise question at issue, that intention is the law and must be given effect."[71] Justice Marshall, in a contemporaneous dissent, remarked that "Chevron's deference requirement . . . was explicitly limited to cases in which congressional intent cannot be discerned through the use of traditional techniques of statutory construction."[72]

Justice Stevens, the author of the Chevron decision, later stated that a "pure question of statutory construction [is] for the courts to decide" by "employing traditional tools of statutory construction."[73] Justice Scalia took issue with this position, asserting that when the statute is not clear, the court must defer to the agency's interpretation.[74] Justice Stevens responded that "your criticism is based on a misreading of the Court's opinion in Chevron. That opinion did not announce a simple black letter rule requiring absolute deference to agencies unless there is clear evidence of congressional intent, and no deference when intent is clear." He stressed that "the more important inquiry is whether the issue is one that Congress intended itself to resolve or whether it is an interstitial area in which Congress intended to delegate lawmaking authority to the agency."[75]

Federal courts vacillated between strong and weak readings of Chevron.[76] The selective application of Chevron indicated that the courts still had an important role to play in statutory interpretation. Chevron should only have been viewed as a judicially imposed prudential concern that was to be applied in a flexible manner.

Recently, the Supreme Court overturned Chevron. The Court, espousing the views of judicial proponents, held Chevron violated the separation of powers by undermining the role of courts to determine the meaning of the law and APA. However, earlier cases relying on Chevron would not be overturned.[77] Even though the federal courts no longer have to defer to the agency's legal interpretations, federal courts still have to utilize the traditional tools of statutory interpretation. Since all of the cases discussed in this book occurred prior to the Court's recent decision, the cases are analyzed under the Chevron doctrine.

1.4.2.2 Judicial Review of Policy Questions

Federal courts are sometimes asked to review the agency's policy decisions in public law litigation. The APA requires the court to ensure that agency actions are not "arbitrary, capricious, an abuse of discretion, or otherwise not in accordance with law." This standard of review supports both the supervisory hard-look doctrine and the deferential soft-glance approach.[78]

The hard-look doctrine requires agencies to consider all relevant factors, justify departures from past practices, furnish detailed explanations of their decision, explain the rejection of alternatives, and show the connections between statutory purposes and regulatory policies. The court examines the agency's action "to satisfy itself that the agency has exercised a reasoned discretion with reasons that do not deviate from or ignore the ascertainable legislative intent."[79] The court "attempts to make sense out of vague delegations, enforce policy declarations discernible there, and . . . reconcile contradictory and contrived legislative histories." The hard-look doctrine "ensures that the agency's decision was a 'reasoned' exercise of discretion and not merely a response to political pressures."[80]

Judicial skeptics have raised objections to the hard-look doctrine. First, a detailed explanation has little substantive effect on the agency's decision. Second, the courts lack the expertise and the political accountability to play a constructive role in the administrative process. Third, the hard-look doctrine ignores the important role of politics in agency decisions. Fourth, judges use the hard-look doctrine to substitute their own policy choices for those of elected officials. Fifth, the hard-look doctrine undermines the legitimacy of agency decisions. Sixth, the hard-look doctrine is not needed because better political alternatives are available. Finally, aggressive judicial review misappropriates limited agency resources.[81]

Judicial proponents point out that skeptics fail to recognize the necessity and value of the hard-look doctrine. Aggressive judicial review "gives those concerned with well-reasoned decision making a lever with which to move those who do not." Only the courts "have the time, the influence, and the freedom from ceremonial and political considerations that are necessary to a thorough, dispassionate and effective review of extremely complex and controversial matters."[82]

The hard-look doctrine advances the separation of powers. Judicial involvement prevents agency capture. The court ensures that the agency's decisions are based on technocratic expertise and conform to the statutory goals. Politics can influence value choice, but only within the constraints of the statute. Judge Leventhal, the father of the doctrine, noted that the hard-look doctrine combines "judicial supervision with a salutary principle of judicial restraint." Agencies and courts together constitute a "partnership in furtherance of the public interest" and the "court is in a real sense part of the total administrative process, and not a hostile stranger to the office of first instance."[83] Judicial review advances the republican ideal of honest disinterested deliberation regarding the public interest. Federal courts continue to endorse and utilize the hard-look doctrine.[84]

1.4.3 Constitutional Questions

Federal courts in some public law litigation have to determine whether government actions are consistent with the Constitution, including the separation of powers and nondelegation doctrine. Judicial skeptics are critical of the court's interfering with the executive and congressional actions; judicial proponents support active judicial efforts to preserve constitutional restrictions.

1.5 Dialogue

Judicial decisions in public law litigation do not end the ongoing institutional dialogue. Judicial decisions affirm or reject executive-legislative deals manifested in the statute; affect the competitive struggle between the legislative and executive branches over policy; impact executive implementation of policy; and structure federal-state relations. Judicial decisions are "political resources" that are "best viewed as the beginning of a political process."[85] Studies demonstrate that Congress is very much aware of judicial interpretations, devotes significant time to analyzing their policy implications, and frequently overrides those interpretations.[86] James Walker and Michael Solomine observed that "the last word in statutory interpretation belongs to no one institutional or individual, which is as it should be."[87]

This chapter provided the underlying theoretical basis of the litigation dealing with the Mexican wolf's reintroduction and recovery, which will be discussed in the subsequent chapters.

Notes

1 Walter A. Rosenbaum, *Environmental Policy and Politics*, 3d ed. (Congressional Quarterly, 1995), 82–7. *Mistretta v. U.S.*, 488 U.S. 361, 380 (1989).

2 Daniel A. Farber and Philip P. Frickey, "Forward: Positive Political Theory in the Nineties," 80 Georgetown Law Journal 457, 462 (1992).

3 Edwin Meese III, "Foreword," in Lee Edwards (ed.), *Bringing Justice to the People* (Heritage, 2004),

4 Burton A. Weisbrod, "Problems of Enhancing the Public Interest: Toward a Model of Government Failures," in Burton Weisbrod, Joel Handler, and Neil Komesar, *Public Interest Law* (University of California, 1978), 30–41

5 William N. Eskridge Jr., "Dynamic Statutory Interpretation," 135 University of Pennsylvania Law Review 1479, 1511–21, 1530–1 (1987).

6 Frank I. Michelman, "Political Markets and Community Self-Determination: Competing Models of Judicial Legitimacy," 53 Indiana Law Journal 145 (1977–8).

7 Henry Hart and Louis Sacks, *The Legal Process: Basic Problems in the Making and Application of Law* (Foundation, 1994), 694–5.

8 Dennis L. Soden, *The Environmental Presidency* (SUNY Press, 1999). Norman J. Vig and Michael E. Kraft (eds.), *Environmental Policy in the 1990s* (Congressional Quarterly, 1997), 120–3.

9 Kenneth Culp Davis, *Administrative Law Treatise*, section 6.15 (West Law, 1978).

10 Cass R. Sunstein, "Factions, Self-Interest, and the APA: Four Lessons since 1946," 72 Virginia Law Review 271, 291–2 (1986). Daniel B. Rodriguez, "The New Legal Process," 77 California Law Review 919, 927 (1989).

11 Richard J. Pierce, "Lujan v. Defenders of Wildlife: Standing as a Judicially Imposed Limit on Legislative Power," 42 Duke Law Journal 1170, 1171–88 (1993).

12 Neil E. Devins, "Appropriations Redux: A Critical Look at Fiscal Year 1988 Continuing Resolution," 1988 Duke Law Journal 389 (1988).

13 Richard Allen Liroff, "Administrative, Judicial, and Natural Systems: Agency Response to the NEPA of 1969," 3 Loyola University of Chicago Law Journal 19, 29–30 (1972).

14 Abram Chayes, "The Role of the Judge in Public Law Litigation," 89 Harvard Law Review 1281 (1976). Abram Chayes, "The Supreme Court 1981 Term, Public Law Litigation and the Burger Court," 96 Harvard Law Review 4, 58 (1982). Bradley Canon and Charles Johnson, *Judicial Policies: Implementation and Impact*, 2d ed. (Congressional Quarterly, 1998), 131–8.

15 Christopher Smith, *Courts, Politics, and the Judicial Process*, 2d ed. (Nelson Hall, 1997), 225, 287.

16 William M. Landes and Richard A. Posner, "The Independent Judiciary in an Interest-Group Perspective," 18 Journal of Law and Economics 875 (1975).

17 Cass R. Sunstein, "Naked Preferences and the Constitution," 84 Columbia Law Review 1689 (1984). Jonathan R. Macey, "Promoting Public-Regarding Legislation Through Statutory Interpretation: An Interest-Group Model," 86 Columbia Law Review 223 (1986).

18 Karen M. Gebbia-Pinetti, "Statutory Interpretation: Democratic Legitimacy and Legal System Values," 21 Seton Hall Legislation Journal 233 (1997).

19 Martin H. Redish and Theodore T. Chung, "Democratic Theory and the Legislative Process: Mourning the Death of Originalism in Statutory Interpretation," 68 Tulane Law Review 803, 805 (1994).

20 T. Alexander Alienikoff, "Updating Statutory Interpretation," 87 Michigan Law Review 20, 23–4 (1988).

21 *INS v. Cardoza-Fonseca*, 107 S. Ct. 1207, 1224 (1987).

22 Nicholas S. Zeppos, "Legislative History and the Interpretation of Statutes," 72 Virginia Law Review 1295, 1299–305 (1990).

23 Cass R. Sunstein, "Interpreting Statutes in the Regulatory State," 103 Harvard Law Review 405, 418–24, 467 (1989).

24 Matthew D. McCubbins, Roger G. Noll, and Barry R. Weingast, "Positive Canons: The Role of Legislative Bargains in Statutory Interpretation," 80 Georgetown Law Journal 705, 738 (1992). Patricia M. Wald, "The D.C. Circuit: Here and Now," 55 George Washington Law Review 718, 727–8 (1987).

25 William N. Eskridge Jr., "The New Textualism," 37 UCLA Law Review 621, 679–80 (1990).

26 William N. Eskridge Jr. and Philip P. Frickey, "Statutory Interpretation as Practical Reasoning," 42 Stanford Law Review 321, 340–1 (1990).

27 Charles Curtis, *It's Your Law* (Harvard University, 1954), 52. Redish and Chung, 813–15.

28 Robert Weisberg, "The Calabresian Judicial Artist: Statutes and the New Legal Process," 35 Stanford Law Review 213, 246–7 (1983).

29 Antonin Scalia, *A Matter of Interpretation, Federal Courts and the Law* (Princeton University, 1997), 29–37.

30 Patricia M. Wald, "The Sizzling Sleeper: The Use of Legislative History in Construing Statutes in the 1988–89 Term of the US Supreme Court," 39 American University Law Review 277, 306–7 (1990).

31 Stephen Breyer, "On the Uses of Legislative History in Interpreting Statutes," 65 Southern California Law Review 845, 848 (1992).

32 Eskridge, "New Textualism," 630–40. Richard C. Ausness, "Federal Preemption of State Products Liability Doctrines," 44 South Carolina Law Review 242–3 (1993).

33 Max Radin, "Statutory Interpretation," 43 Harvard Law Review 863 (1929–30). Eskridge, "New Textualism," 643–50.

34 Reed Dickerson, *The Interpretation and Application of Statutes* (Little Brown, 1975), 68–79, 85–6.

35 Stephen Breyer, *Reading the Constitution* (Simon & Schuster, 2024), 88.

36 Richard A. Posner, "The Jurisprudence of Skepticism," 86 Michigan Law Review 827, 848–52 (1988).

37 Hart and Sacks, 1124. Vincent Wellman, "Dworkin and the Legal Process Tradition," 29 Arizona Law Review 413, 463 (1987).

38 William N. Eskridge Jr. and Philip P. Frickey, "Legislation Scholarship and Pedagogy in the Post-Legal Process Era," 48 University of Pittsburgh Law Review 691, 704–9 (1987).

39 Hart and Sacks, 1374–80.

40 Breyer, *Reading the Constitution*, 105–9.

41 Macey, 261–7. Farber and Frickey, 475.

42 Stephen Breyer, "Judicial Review of Questions of Law and Policy," 38 Administrative Law Review 363, 365–8 (1986).

43 Colin S. Diver, "Statutory Interpretation in the Administrative State," 133 University of Pennsylvania Law Review 549, 568–92 (1985).

44 Donald Horowitz, *The Courts and Social Policy* (Brookings Institution, 1977).

45 R. Shep Melnick, "Administrative Law and Bureaucratic Reality," 44 Administrative Law Review 245 (1992). Robert A. Kagan, "Adversarial Legalism and American Government," 10 Journal of Policy Analysis and Management 369 (1991).

46 Martin Shapiro, "Judicial Activism," in Seymour Lipset, *The Third Century: America as a Post-Industrial Society* (Hoover Institute, 1979), 109–133.

47 *Associated Data Processing Service Organizations v. Camp*, 397 U.S. 150 (1970).

48 Michael S. Greve, *The Demise of Environmentalism in American Law* (American Enterprise Institute, 1996), 42–63. Jeremy Rabkin, *Judicial Compulsions: How Public*

Law Distorts Public Policy (Basic Books, 1989), 19–31. R. Shep Melnick, *Regulation Through the Courts* (Brookings Institution, 1983).

49 Victor Aubert, "Competition and Dissensus: Two Forms of Conflict Resolution," 7 Journal of Conflict Resolution 26 (1963). Lon Fuller, "The Forms and Limits of Adjudication," 92 Harvard Law Review 92 (1978).

50 *Calvert's Cliff Coordinating Committee v. AEC*, 449 F.2d 1109, 1115 (D.C. Cir. 1971).

51 Abner J. Mikva, "How Should the Courts Treat Administrative Agencies?," 36 American University Law Review 1, 3 (1986). Cass R. Sunstein, "What's Standing After Lujan?" 91 Michigan Law Review 163, 212 (1993).

52 5 U.S.C. §701, §706 (1988). Cass R. Sunstein, "Law and Administration After Chevron," 90 Columbia Law Review 2071, 2080–1 (1990).

53 Cass R. Sunstein, "On the Costs and Benefits of Aggressive Judicial Review of Agency Action," 1989 Duke Law Journal 522, 525 (1989).

54 Louis Jaffe, *Judicial Control of Administrative Action* (Little Brown, 1965), 320, 325.

55 Chayes, "Role of the Judge," 1307–8. Richard A.L. Gambitta, Marlynn L. May, and James C. Foster, *Governing Through Courts* (Sage, 1981).

56 16 U.S.C. 1540 (g).

57 *Sierra Club v. Morton*, 405 U.S. 727 (1972).

58 *U.S. v. Students Challenging Regulatory Agency Procedures*, 412 U.S. 699 (1973). *Duke Power v. Carolina Study Group*, 438 U.S. 59 (1978).

59 Gene R. Nichol Jr., "Justice Scalia, Standing, and Public Law Litigation," 42 Duke Law Journal 1141 (1993).

60 Richard Pierce, "The Role of the Judiciary in Implementing an Agency Theory of Government," 64 New York University Law Review 1239, 1283–4 (1989).

61 Henry P. Monaghan, "Marbury and the Administrative State," 83 Columbia Law Review 1, 12 (1983). 5 U.S. (Cranch) 137, 177–8 (1803).

62 Maureen B. Callahan, "Must Federal Courts Defer to Agency Interpretations of Statutes?," 1991 Wisconsin Law Review 1275, 1278–9 (1991). Thomas W. Merrill, "Judicial Deference to Executive Precedent," 101 Yale Law Journal 969, 972–3 (1992).

63 467 U.S. 837, 842–5 (1984).

64 467 US, 865–6.

65 Frank B. Cross, "Shattering the Fragile Case for Judicial Review of Rulemaking," 85 Virginia Law Review 1243, 1249–50 (1999).

66 Kenneth W. Starr, "Judicial Review in the Post-Chevron Era," 3 Yale Journal on Regulation 283, 310–11 (1986).

67 Cass R. Sunstein, "Constitutionalism After the New Deal," 101 Harvard Law Review 421, 466–9, 488–99 (1987).

68 Cynthia R. Farina, "Statutory Interpretation and the Balance of Power in the Administrative State," 89 Columbia Law Review 452, 478–88 (1989).

69 Breyer, "Judicial Review of Questions of Law and Policy," 379. *Arlington Texas v. FCC*, 133 S. Ct. 1863, 1876–7 (2013).

70 Kenneth W. Starr, Cass R. Sunstein, Richard K. Willard, and Alan B. Morrison, "Judicial Review of Administrative Action in a Conservative Era," 39 Administrative Law Review 353, 367 (1987).

71 467 US, 843 n.9.

72 *Chemical Manufacturers Assoc. v. NRDC*, 470 U.S. 116 n.15 (1985).

73 *INS v. Cardozo-Fonseca*, 480 U.S. 421, 446 (1987). Callahan, 1295–9.

74 480 US, 452–6.

75 Bernard Schwartz, "Shooting the Piano Player? Justice Scalia and Administrative Law," 47 Administrative Law Review 1, 47–8 (1995).

76 Jason J. Czarnezi, "An Empirical Investigation of Judicial Decisionmaking, Statutory Interpretations, and the Chevron Doctrine in Environmental Law," 79 University of Colorado Law Review 767, 782–3 (2008).

77 *Loper Bright Enterprises v. Raimondo*, No. 22–451 (June 28, 2024).

78 William H. Rodgers, *Environmental Law*, 2d ed. (West, 1994), 90–111.

79 *Greater Boston Television Corp. v. FCC*, 444 F.2d 841, 850–1 (D.C. Cir. 1970).

80 William H. Rodgers, "Judicial Review of Risk Assessments," 11 Environmental Law 301, 317 (1981). Cass R. Sunstein, "Deregulation and the Hard-Look Doctrine," 1983 Supreme Court Review 177, 181–4 (1983).

81 Joseph L. Sax, "The (Unhappy) Truth About NEPA," 26 Oklahoma Law Review 239, 247–8 (1973). Peter L. Strauss, "Considering Political Alternatives to 'Hard Look' Review," 1989 Duke Law Journal 538, 548 (1989). Martin Shapiro, *Who Guards the Guardians? Judicial Control of Administration* (University of Georgia, 1988), 151–6.

82 William F. Pedersen Jr., "Formal Records and Informal Rulemaking," 85 Yale Law Journal 38, 60 (1975).

83 Harold Leventhal, "Environmental Decisionmaking and the Role of the Courts," 122 University of Pennsylvania Law Review 509, 511–12 (1974).

84 Cass R. Sunstein, "Interest Groups in American Public Law," 38 Stanford Law Review 29 (1985). *Motor Vehicle Manufacturers' Assoc. v. State Farm Mutual*, 463 U.S. 29 (1983).

85 Stuart Sheingold, *The Politics of Rights: Lawyers, Public Policy and Political Change* (Yale University, 1974), 85.

86 Thomas R. Marshall, "Policymaking and the Modern Court: When Do Supreme Court Rulings Prevail?," 42 Western Political Quarterly 493, 501 (1989). William N. Eskridge Jr., "Overriding Supreme Court Statutory Interpretation Decisions," 101 Yale Law Journal 331, 334–8 (1991).

87 James L. Walker and Michael E. Solomine, "The Next Word: Congressional Response to Supreme Court Statutory Decisions," 65 Temple Law Review 425, 453 (1992).

2

The Extermination and Resurrection of the Wolf

Lobo's fate was tied to changes in public land management. The livestock industry expanded in the southwest. State bounties were enacted to stop wolf depredation, but had mixed success. The federal government began to set aside and lease public lands. Lessees demanded federal predator control, which nearly eliminated the Mexican wolf from the southwest. Changes in public land policy, continued scientific criticism of predator control, concern for species extinction, the emergence of the environmental movement in the 1960s and 1970s coupled with major political and legal changes facilitated lobo's return. The enactment of the Endangered Species Act (ESA) in 1973 provided the mandate and means for the reintroduction and recovery of the Mexican gray wolf. At the same time, the change in public land management policies from resource development to environmental protection led to a backlash, which was evidenced in the Sagebrush Rebellion, the Wise Use, and the County Supremacy Movements. This chapter examines the extermination and resurrection of the Mexican wolf in the context of changes in public land management.

2.1 Extermination

2.1.1 State Bounties

Cattle ranching dominated the southwest in the nineteenth century. Large unregulated herds faced threats from Apaches and wolves. After the Civil War, the federal army moved into the southwest and eliminated the Apache threat. In the 1870s and 1880s, railroad expansion allowed western livestock to move east. By the 1890s, the open range was gone and cattle ranching became a competitive business. The number of cattle in Arizona increased from 50,000 mid-century to an estimated 1.5 million in 1891.[1] Sheep ranching also expanded on the range. In 1893, the livestock industry in the southwest collapsed as a result of drought and overgrazing. Every animal became crucial, so ranchers could not afford to lose livestock to predators. Ranchers pushed for bounties that were established at the state, county, and local levels. Ranchers argued that "each head of livestock killed by wolves reduced the number that could be taxed."[2]

Bounty systems were implemented. The 1893 Arizona-New Mexico Territorial Bounty Act allowed counties to appropriate money for bounties on wolves, coyotes, and other predators. In 1909, the New Mexico territorial legislature enacted a bounty act. Each county was authorized to levy an annual tax not to exceed one mill on the assessed value of all the county's taxable property to maintain the "Wild Animal Bounty Fund." The bounty was $15 per lobo wolf and $2 per coyote. Bounty hunters were required to make a claim within ninety days after the killing by filing an affidavit before the county clerk, swearing the animal was killed in the county. The claimant had to present the scalp of the dead animal or the entire hide of the lobo to collect.[3]

The bounty system worked well in the short term. Counties assessed the tax. Bounties were paid and the targeted species decreased. Some counties then decided not to assess the tax, so bounty hunters took the animals to the neighboring counties paying bounties. Recognizing the free rider problem, the neighboring counties stopped assessing the tax. Inadequate funding placed the bounties in arrears.

The New Mexico Cattle Growers Association (NMCGA) and other private parties decided to supplement the public funds. District associations offered their own bounties, providing an additional $25 for wolves and $10 for every wolf pup killed on the range. Financial problems quickly emerged. The associations required hunters to assign their county bounties to them, but these were five to six years in arrears. The NMCGA began to question the efficacy of bounties for controlling predators. Whenever predators became scarce in one area, bounty hunters moved to another area. The predators' high reproductive capability allowed them to repopulate the prior area. Bounties also provided incentives for hunters to maintain a supply of predators to sustain their livelihood. In 1916, the NMCGA stopped its bounty program and directed its efforts toward securing federal and state funds for predator control.[4]

In the late nineteenth century, hunters joined the livestock industry, calling for predator control. The commercial market on game animals emerged. New technology, the refrigerated box car, allowed hunters to establish a national market for wild game.[5] At the same time, national organizations were formed, such as the Boone and Crockett Club, which supported the preservation and regulation of wildlife to supply game for hunters. After Arizona and New Mexico became states in 1912, sportsmen lobbied for the establishment of protective game associations, which were formed in New Mexico in 1914 and Arizona in 1923. Commercial hunters and sportsmen joined the ranchers and demanded that the federal government kill Mexican wolves to protect the supply of game.[6]

2.1.2 Federal Predator Control

The federal government today owns most of the land west of the Mississippi, including 43 percent of Arizona and 33 percent of New Mexico. At the end of the nineteenth century, the federal government began to withdraw and set aside public lands for defined purposes, including the establishment of national forests, national parks, and wildlife refuges. Federal administrative systems, including the U.S. Forest Service (FS) and the National Park Service (NPS), were established to manage and control over 625 million acres, 96 percent of the public domain.

The first major set aside was the national forest land in 1897. The FS charged fees for grazing in the national forests. Livestock interests argued it was not fair to charge a fee for grazing on public land populated by predators unless predator protection was provided. The FS, seeking to develop a constituency for federal retention of the national forests, supported predator control. The FS began to kill predators in the southwest. In 1908, 232 wolves were killed in national forests in New Mexico and 127 wolves were killed in national forests in Arizona.[7] The NPS soon followed the practice.

The FS efforts led to the expansion of the U.S. Biological Survey (BS) in the Department of Agriculture, which had been established in 1885 to study the effects of birds and mammals on agriculture. Initially, the BS was an advisory service. In 1905, the BS was renamed the Bureau of Biological Survey (BBS) and was tasked with the mission of locating wolves for the FS to kill. In 1907, the BBS noted that wolves and coyotes levied a heavy tax on stock, but predator losses could be prevented by a carefully planned eradication effort in wolf-infested country.

The BBS published several reports between 1907 and 1909 regarding federal predator control. Vernon Bailey's 1907 bulletin, "Wolves in Relation to Stock, Game, and National Forest Reserves," noted the tremendous loss of livestock due to wolf depredation. The bulletin provided hunters, trappers, forest rangers, and ranchers directions for trapping, poisoning, and hunting wolves and finding wolf dens. Bailey's 1909 report, "Key to Animals on Which Wolf and Coyote Bounties Are Often Paid," provided advice to states and counties on how to reduce fraud in bounty systems. Statistics showed that more than $100,000 annually was being paid out in bounties. This did not include local bounties paid by livestock associations.[8] The relationship between the FS and the BBS increased.

The onset of the First World War increased the demand for US meat to feed the troops. The livestock industry put pressure on western congresspersons, who held influential positions on congressional committees, for federal predator control. Congress granted the BBS authority over predator control and appropriated $125,000 for the BBS to conduct predator control on public lands and national forests in 2014 and 2015. The outbreak of rabies in 1915, which killed livestock, was blamed on wolves and other predators. In 2016, BBS appropriations were increased by $75,000 to stop rabies. By 2017, $250,000 per year was being spent on predator control.[9] The BBS was transformed from an advisory to a service agency.

By the end of 1916, the Predator and Rodent Control (PARC) branch of the BBS was organized into eight districts. Stokley Ligon, the head of the combined New Mexico/Arizona district, determined that 300 Mexican wolves were present in New Mexico. He estimated that each wolf killed three cows per month, resulting in the death of 10,800 cows per year. At $30 per head, wolf depredation amounted to $324,000 per year.[10] The BBS hired professional hunters and trappers, who utilized steel traps, poison baits, denning, shooting, and roping to kill predators. Poisoning with strychnine provided the easiest and quickest means. Bounties were not offered. Predator fur was sold at auction and the funds deposited into the Treasury. There was constant friction between private trappers, who sought bounties, and salaried government hunters. The PARC took sixty-eight adult wolves and seventeen pups in FY 1916–17, mostly in New Mexico.[11]

In 1917, the NMCGA, the New Mexico Wool Growers (NMWG), and the State Game Protective Association demanded the secretary of agriculture allocate more money for predator control. They claimed that predators generated $6 million per year in livestock losses, which translated into the loss of 50 million pounds of meat. The loss of livestock was detrimental to the war effort. The secretary of agriculture established a federal matching fund of $25,000 for the BBS predator control.[12]

The BBS was captured by the livestock industry. After the BBS became a service agency, its clientele, ranchers and woolgrowers, gained more influence because they paid the bills. The demand for federal predator control exceeded the available funding, so the cost was shifted over to the livestock industry. In 1918, the BBS began cooperative programs with states, counties, and livestock associations. The BBS investigated complaints and provided hunters and equipment, while its clientele funded the effort through a head tax on livestock in the affected areas. New Mexico was split into two districts in 1918. Mark Musgrave headed the Arizona District. Ligon, who remained in control in New Mexico, declared that "the big wolves have been so reduced in numbers in New Mexico and Arizona that they no longer confront us as a serious menace."[13]

In 1919, Musgrave estimated that Arizona experienced $1.5 million in livestock loss, while Ligon claimed $265,000 in livestock lost due to wolf depredation. Nevertheless, Ligon's summary report outlining activities from February 1918 to December 1919 stated that depredations by wolves had been reduced by three-fourths since November 1917. He repeated that wolves were no longer a serious threat, but warned that trapping the remaining wolves was going to be difficult and time-consuming. The PARC took seventy-eight wolves in FY 1918–19.[14]

Several states decided to abandon bounties and rely on federal predator control. Arizona and New Mexico cooperated with the PARC. In 1919, Arizona enacted a statute mandating that state funds had to match the federal expenditures spent on predatory animal control in the state. In 1919, the New Mexico legislature, seeking to make the federal-state program permanent, enacted a statute that established a cooperative program with the BBS for the destruction of predatory wild animals to protect livestock and improve range conditions. In 1923, New Mexico repealed the 1909 bounty act and increased the predatory animal fund by $25,000. The state wanted the federal government to match its $50,000 annual commitment, but the federal government never spent more than $32,000–33,000 annually during the decade.[15] In FY 1919–20, 110 adult wolves and at least twenty-one young wolves were taken by the PARC.

The PARC's "cooperation" mandate encouraged livestock industry participation. The NMCGA and the NMWG acted as advisers to the BBS. Regional stockmen associations sprang up to deal with predator control. The regional associations elected managers, levied assessments, and paid the salaries of hunters, who coordinated their activities with the BBS. By 1922, there were forty associations advising the BBS. By the mid-1920s, the PARC received one-quarter of its funding from its clients. The PARC took eighty-six wolves between 1921 and 1922.[16]

There was a rapid reduction in the wolf population from 1915 through the 1920s. In 1924, Ligon stated that "a survey of the wolf situation in the state indicates that these, the greatest cattle takers, are no longer a real menace." The PARC took sixty-three wolves in 1924, mostly along the border. By 1925, the wolf ceased to be a major predator in the southwest. The PARC focused its efforts on stopping wolves coming in from Mexico across the San Madre Mountains, principally through Hidalgo County, New Mexico. This promoted interest in Hidalgo County to create a "wolf proof" fence along the border, foreshadowing future events.[17]

In 1928, the secretary of agriculture was called on to investigate and report on "the feasibility of a ten-year cooperative program . . . for the purposes of eradication, suppression, or bringing under control of predatory animals within the United States and estimated costs thereof." The secretary reported that such an effort would be possible over the next ten years and would save $10 for every $1 spent. Complete eradication was not possible, but the lobo could be brought under control. The cost of federal predator control over the past thirteen years averaged $432,042 per year. The annual price tag for the new ten-year program was $1,378,700, or 243 percent above the amount apportioned for fiscal year 1929.[18]

2.1.3 Scientific Skepticism

Scientists questioned federal predator control. In December 1918, scientists formed the American Society of Mammalogy (ASM), which advocated for predator conservation. Objections were raised regarding the federal extermination of predators at the ASM meeting in 1924. In response, the BBS created a scientific committee in 1924 to study federal predator control. Its 1928 report noted that predators should be restricted to national parks and wilderness areas. Many scientists disagreed with the report's findings. The critics accused the BBS of cooperating with the western livestock industry and state governments to conduct a politically motivated extermination campaign that was based on false data and bad science. Ranchers and state officials exaggerated their livestock losses. Predators did not pose a serious threat to livestock. Predator control policy should be replaced by a "system of intelligent controls" adapted to the specific needs of different western regions.[19]

The debate regarding predator control occurred against the backdrop of deer irruptions in Kaibab National Forest and Gila National Forest (GNF) in the 1920s, which demonstrated the harmful ecological consequences of federal predator control. In 1906, President Roosevelt designated the Kaibab portion of Grand Canyon Forest Reserve as a national game preserve, which also supported 20,000 sheep and cattle. Game hunting was outlawed. Government hunters proceeded to kill all the predators in the area. By 1924, over 6,000 wolves, coyotes, and mountain lions had been killed. With the absence of predators, the deer population exploded from 4,000 to over 100,000, denuding vegetation and causing an ecological disaster. The secretary of agriculture ordered the killing of deer to stop the overgrazing. Colorado brought suit alleging a violation of the state's gaming law, but the Supreme Court upheld the federal government's authority "to protect its land and property." Similar events occurred in the GNF.[20]

In 1930, the ASM protested the killing of wildlife and questioned the rationale for killing predators. In response, Congress delayed the BBS appropriation. The PARC pleaded with the ASM to withdraw opposition and promised to target and end its killings. The ASM sent a letter to Congress limiting its opposition solely to predator control. Congress then released the PARC's appropriation. Despite this ostensible new start, the PARC returned to its prior practice and continued to indiscriminately kill predators. Ironically, the PARC used the Society's letter to justify predator control.[21]

Congress did not intend to stop the war against predators. In 1931, Congress enacted the Animal Damage Control Act (ADCA), which provided statutory authority to the PARC. The secretary of agriculture was authorized to investigate the best methods to control and eventually eradicate predators that interfered with human activities on all public and private lands. The ADCA authorized $1 million per year from 1932 to 1941, but no funds were appropriated because of the depression and preparations for the Second World War. Critics pointed out the ADCA "spells out no central aim or purpose; its implicit premise is that all 'injurious' species should be destroyed."[22]

The enactment of the ADCA foreshadowed the closing of the public domain. The Taylor Grazing Act, enacted in 1934, established grazing districts on public lands. The secretary of interior was authorized to regulate grazing, issue permits for grazing, and charge a grazing fee. Ranchers soon dominated the program through advisory boards. The Grazing Service, now the Bureau of Land Management (BLM), supported predator control on public lands.[23]

The scientific management of wildlife emerged in the 1930s. Aldo Leopold introduced the concepts of ecosystem management and carrying capacity as wildlife management tools.[24] Scientists continued to criticize federal predator control. Leopold recommended that "no predatory species should be exterminated over large areas" and that "rare predatory species . . . should not be subject to control." Sigurd Olsen's 1938 study of wolves and coyotes in Superior National Forest demonstrated that predators were necessary to maintain the ecological balance in the wilderness. Olsen noted that the wolf "is an integral part of the wilderness community, the destruction of which would destroy the fine balance between related forms." Nevertheless, Olsen recognized that "the politics of wolf preservation and the science of studying wolves is more vicious and complicated than any wolf pack I've had the pleasure of studying."[25]

In response, the NPS changed its policy regarding predator control. The NPS declared that "no native predator shall be destroyed on account of its normal utilization of any other park animal, excepting if that animal is in immediate danger of extermination, and then only if the predator is not itself a vanishing form."[26] The Departments of Agriculture and Interior did not follow this policy but continued predator control on public lands.

2.1.4 Continued Persecution

In 1939, the BBS moved to the Department of Interior, where the livestock industry exerted even greater control over the program. In 1940, the BBS was combined with the Bureau of Fisheries to form the U.S. Fish and Wildlife Service (FWS). By this time, only a few Mexican wolves remained in the United States. Yet, predator control expanded, with the coyote becoming public enemy number one. The PARC increasingly relied on poisons to kill predators. Two new toxicants thallium sulfate and compound 1080 (sodium fluoroacetate), as well as the "coyote getter," which consisted of an exploding cyanide capsule hidden in material attractive to animals, were introduced in the 1940s. Federal efforts were so successful that Stanley Young, in 1944, concluded that "the wolf has been definitely brought under control and presents a very minor problem, except in limited areas in the United States."[27] Congress doubled the PARC budget from 1940 to 1950, and again by 1970.[28] From 1937 to 1970, the PARC killed 1,574 Mexican wolves.

Mexico also began wolf eradication in response to livestock losses. In the 1930s and 1940s, Mexican ranchers employed the same techniques perfected in the United States, such as traps and poisons, to reduce the wolf population. In the 1950s, the FWS and the Pan American Sanitary Bureau utilized compound 1080 along the border, ostensibly to prevent the outbreak of rabies that flared up in 1945. The program was initiated and continued at the insistence of the livestock industry. By the 1970s, there was limited wolf migration across the border from the few Mexican wolves remaining in the Sierra Madre Mountains.[29]

2.2 Post–Second World War Management Changes

Public attitudes toward public lands, wildlife, and the wolf changed after the Second World War. With growing economic prosperity, citizens took advantage of recreational opportunities. Public lands and wildlife were no longer viewed exclusively as a source of resources, but their scientific, recreational, commercial, and ecological values were recognized. Arthur Leopold, father of conservation biology, proposed the "land ethic" for public land management, which aimed to preserve biodiversity and ecosystems. According to Leopold, "a thing is right when it tends to preserve the integrity, stability, and beauty of the biotic community. It is wrong when it tends otherwise."[30]

The perception of the wolf was beginning to change. Leopold reflected this new view when discussing his killing of a Mexican wolf in the early 1900s. He stated:

> We reached the old wolf in time to watch the fierce green fire dying in her eyes. I realized then, and have known ever since, that there was something new to me in those eyes—something known only to her and to the mountain. I was young then, and full of trigger itch; I thought that because fewer wolves meant more deer, that no wolves would mean hunter's paradise. But after seeing the green fire die, I sensed that neither the wolf nor the mountain agreed with such a view.[31]

There were also changes in public land management that stressed greater protection for wildlife. There was growing recognition that "interest group liberalism" had failed. Congress delegated broad administrative authority to public land agencies, including FS and BLM, which became captured by the regulated commodity industries, particularly the timber and livestock industries. Public land policy was being conducted and made in a closed administrative process through subgovernments, or "iron triangles of government," consisting of interest groups, congressional committees, and executive agencies. Public land policy did not serve the public interest, but private interests.[32]

In the early 1960s, there was a political effort to move from dominant-use toward multiple-use public land management. New statutes were enacted that expanded federal agency mandates and provided for greater interest group participation in the administrative process. Congress created an "interest representation scheme," which was based on the premise that increased pluralism in the administrative process would create a more balanced approach to realize the public interest.[33]

These multiple-use statutes did not remedy agency capture. There was little change in the direction of policy because the statutes did not establish priorities. Administrators, working under the same political constraints, were only required to consider the enumerated factors but had discretion regarding their importance.[34] Conservation groups did achieve some progress regarding the protection of species habitat. The Land and Water Conservation Act of 1964 provided federal funds for the acquisition of land for the preservation of species of fish or wildlife species threatened with extinction. The Wilderness Act of 1964 and National Wildlife Refuge System Act of 1966 set aside public lands, which benefited wildlife.[35]

2.3 Emergence of the Environmental Movement

The modern environmental movement arose in the 1960s and 1970s with the baby boom generation, which was raised during economically affluent times and manifested post-industrial values. Faith in science and technology and unfettered economic growth were questioned. There was a recognition of the collective nature of social problems and demand by affected groups and citizens for government action. Events in the 1960s, such as the Santa Barbara oil spill and the Cuyahoga River fire, provided the impetus for the environmental movement.[36]

There was an explosion of congressional and administrative activity in the 1960s and 1970s. New statutes were enacted with bipartisan support that established broad environmental mandates for all federal agencies and provided for public participation. New congressional committees and subcommittees were established with jurisdiction over environmental statutes, which provided public interest groups with points of access into the political process. New executive agencies were formed to provide advice and implement environmental statutes. New environmental groups emerged, which, along with established conservation groups, promoted nonconsumptive values in Congress and executive agencies.[37] Environmental groups were able to participate in the iron triangles of government.

Federal courts became the forum for resolving policy conflicts. Public interest law firms, which were funded by foundations and member contributions, emerged. Citizen suit provisions in the environmental statutes and the Administrative Procedures Act (APA) allowed environmental groups to act as private attorney generals enforcing statutory mandates. Federal courts recognized broad-based environmental injuries and supported environmental interest group standing. Environmental groups, which learned from earlier NAACP and ACLU litigation efforts, brought suits challenging federal government action. Federal courts scrutinized agency decisions to ensure compliance with myriad environmental statutes. Public law litigation emerged.[38]

2.3.1 Endangered Species Act

The emergence of the environmental movement in the 1960s produced a change in the public perception of wildlife. New federal statutes protected wildlife and elevated the status of wildlife in the management of national forests and public lands. The federal government's war on predators was criticized on biological, economic, and political grounds. The image of the wolf changed from a wanton killer to a noble hunter. The wolf became a symbol of the wilderness that must be preserved.[39]

Species destruction and preservation were among the major concerns of the emerging environmental movement. There had been some early concern with species extinction. Aldo Leopold in 1936 wrote "Threatened Species," which argued that the preservation of species should be an important goal of the conservation movement. The NPS and FWS in 1942 published "Fading Trails: The Story of Endangered American Wildlife," which argued for the preservation of wildlife species, including predators, "to help maintain the elusive balance between all living things."[40]

In November 1959, Arthur H. Carhart published "Poisons—the Creeping Killer," which criticized the BBS's use of poisons, specifically compound 1080. This foreshadowed Rachael Carson's *Silent Spring* in 1962, which pointed out the environmental dangers of pesticides. Federal predator control was criticized for using poisons, which indiscriminately killed wildlife and upset the ecological balance and biodiversity. The seeds of doubt sown in the 1930s regarding poison blossomed in the 1960s. Nevertheless, the PARC was still eradicating wildlife.[41]

Facing congressional pressure, Secretary of Interior Stewart Udall created the Leopold Commission to examine federal predator control. The Leopold Report in 1964 concluded that federal predator control was no longer a balanced component of animal husbandry. The PARC was killing more predators than necessary and needed proper management. It had become a semi-autonomous agency whose role had expanded beyond its need. It was servicing the livestock and agriculture industry, which paid its bills. A reassessment of the PARC's goals and implementation was necessary to reflect the public interest. The PARC ignored the recommendations, which were opposed by the livestock industry.[42]

After the Leopold Report, the PARC was transferred to the Department of Interior and became the Division of Wildlife Services (WLS) in 1965. The WLS was responsible for pesticide appraisal and monitoring, as well as wildlife enhancement. The WLS declared that all wildlife was valuable, so "government policy should be one of husbandry of all forms of wildlife." Furthermore, predator control should be limited to "troublesome species, preferably troublesome individuals, and only to localities where substantial damage or danger exists." Even though the WLS was formed both for conservation and control of wildlife, most of its attention focused on control (90 percent killing and 10 percent conservation). The WLS was still killing the few Mexican wolves migrating from Mexico to satisfy its principal constituent, the livestock industry.[43]

Federal predator control became intertwined with species extinction. The Department of Interior in 1964 created the Committee on Rare and Endangered Wildlife Species. The committee published the Redbook in 1966, which provided a preliminary list of rare and endangered species. The Redbook was not legally enforceable but turned attention to species extinction. Scientists working on the Redbook became advocates for species preservation legislation. In 1966, Representative John Dingell (D. Mich.) held hearings on the management of wildlife resources. The hearings provided a forum for groups opposed to federal predator control. These efforts led to the enactment of the Endangered Species Preservation Act (ESPA) in 1966. The secretary of interior was directed to carry out a program to conserve and encourage the growth and replacement of selective species of native fish and wildlife threatened with extinction. The ESPA did not engender the same degree of hostility as the predator control bills, so it did not generate much opposition.

The ESPA was strengthened by the enactment of the Endangered Species Conservation Act (ESCA) in 1969, which protected species threatened with worldwide extinction, including the Mexican wolf. The ESCA established a list of fish and wildlife threatened with extinction and prohibited their importation. In addition, the buying or selling of any animal taken in violation of state or federal law was prohibited. The funding authorization was also increased.[44]

The Public Land Law Commission Report in 1970 fueled further opposition to federal predator control. The report concluded that "federal predator control programs should be eliminated or reduced on Federal public lands." Federal predator control programs "have substantially reduced and in some cases virtually eliminated certain species that are natural predators." While these programs may have been of some benefit to livestock operators, "they have upset important natural mechanisms for the population control of other species." This caused an increase in the ungulate populations "in some localities to levels far above the capacity of the natural habitat to support them . . . resulting in habitat destruction and starvation."[45]

Books and articles continued to focus public attention on the chemical warfare against wildlife. The federal poisoning of predators became a national issue. In 1971, the Council on Environmental Quality and the Department of Interior sponsored a joint study on federal predator control. The Cain Commission report, echoing the findings in the Leopold Report, recognized that the federal predator control program "contains a high degree of built-in resistance to change." The public-private funding scheme "maintains a continuity of purpose in promoting the private interest of livestock growers, especially in the western rangeland states." The livestock industry's financial contribution "serves as a gyroscope for the bureaucratic machinery pointed towards the familiar goal of the general reduction of predator populations, with little attention to the effects of this on the native wildlife fauna."[46]

The Cain Commission determined that predator control had little impact on predator populations. Livestock depredation was not constant and difficult to measure. Although the overall losses were slight, some ranchers suffered an inordinate burden. Heavy livestock losses were isolated and attributed to individual predators, who had acquired a taste for livestock. Furthermore, the resulting costs of ecological disruptions outweighed the benefits of federal predator control. The Commission made several specific recommendations and called for the end of nonselective control methods and an emphasis on individual animals that caused losses.[47]

The Cain Commission report, coupled with numerous eagle killings, prompted President Nixon to issue Executive Order 11,643 in 1972, which restricted the use of toxins for predator control on federal lands. President Nixon stated, "Americas today set high value on the preservation of wildlife. The old notion that 'the only good predator is a dead one' is no longer acceptable."[48] Predator control should not be an end in itself, but must be more selective and ecologically friendly. The EPA halted the interstate shipment of all pesticides registered for use against predators and suspended and canceled the registration for nineteen products containing thallium sulfate, sodium cyanide, strychnine, and compound 1080.[49]

President Nixon proposed a new revised Animal Damage Control Act, which would end federal predator control. Supporters of the bill argued that predators fill an important niche in the ecosystem and play an important role to keep prey in check, preventing overgrazing. They provide important aesthetic value for people and enhance the tourist industry. Predators are intimately linked to the idea of the West and wilderness. Hunters value predators as trophy game. Humans have an ethical and moral obligation not to destroy species, but to share the land with them. Supporters also pointed out that Congress failed to change the 1931 law because the relevant congressional committees dominated by Western interests made any changes impossible.[50]

The livestock and agricultural industries vigorously opposed the bill. They demanded more predator control and the transfer of the WLS back to the Department of Agriculture (accomplished in 1986). The bill passed the House, but failed in the Senate. Other bills, which prohibited the poisoning of predators, provided for the study of predator control methods, established a national policy on predators, and prohibited the killing of endangered species, were also defeated.[51]

The secretary of interior's narrow interpretation of the ESCA precluded its proper implementation. In 1972, President Nixon acknowledged that "even the most recent act to protect endangered species . . . simply does not provide the kind of management tools needed to act early enough to save a vanishing species."[52] Several preceding events facilitated the enactment of the ESA. The Golden Eagle Protection Act of 1972 expanded the scope of federal protection under the Bald Eagle Protection Act of 1940. The Wild Free-Roaming Horses and Burros Act protected wild horses and burros on public lands. The Marine Mammal Protection Act extended federal protection over marine mammals. The United States signed and ratified the Convention on International Trade in Endangered Species (CITES), which required implementing legislation. A wolf specialist group in the International Union for the Conservation of Nature provided a blueprint for global wolf conservation, declaring that wolves, like other wildlife, "have a right to exist in a wild state," one that "derives from the right of all living creatures to coexist with man as part of natural ecosystems."[53]

The livestock industry fought hard to oppose President Nixon's Animal Damage Control Act, so it had little political will to fight expanded species protection. The ESA was enacted in 1973. It protects endangered and threatened species and requires the federal government to restore endangered species. President Nixon's signing statement declared: "Nothing is more priceless and more-worthy of preservation than the rich array of animal life with which our country has been blessed. It is a many-faceted treasure, or value to scholars, scientists, and nature lovers alike, and it forms a vital part of the heritage we all share as Americans."[54]

The ESA prevents federal, state, and private actions that pose a danger to endangered and threatened species. The ESA acknowledges that endangered and threatened species provide "aesthetic, ecological, educational, historical, recreational and scientific value to the Nation" and "provides a means whereby the ecosystems, upon which endangered species and threatened species depend may be conserved."[55] The Supreme Court noted that Congress intended "to halt and reverse the trend towards species extinction, whatever the cost." The legislative history indicated that federal agencies must "afford first priority to the declared national policy of saving endangered species," which has "priority over [their] primary missions." Endangered species must be protected because "it is in the best interests of mankind to minimize the loss of genetic variations," which are potential resources. Endangered species provide "keys to puzzles which we cannot solve and may provide answers to questions which we have not yet learned to ask." The courts must act with "institutionalized caution" when reviewing ESA cases.[56] The ESA established the mandate and means to restore and protect the lobo.

2.4 Conservative Backlash

The new policy of environmental protection angered the resource and commodity interests which had dominated western public land management. New statutes were enacted that established environmental mandates, planning processes, public participation, and access to federal courts. The traditional benefactors of federal land management saw their access to western lands and water being threatened by a new set of policy actors with a different agenda. The enactment of the Federal Land Policy and Management Act in 1976 (FLMPA), which ended the disposition of most public lands and established permanent federal retention, was the final straw. The resource and commodity industries joined with disaffected western state governments to demand the federal government surrender its control over public lands. This federal-state struggle became known as the Sagebrush Rebellion or "the Great Terrain Robbery."[57]

The Sagebrush Rebellion began in 1978 when western states and counties decided to challenge federal ownership of public lands. Different strategies were pursued. Bills were introduced into Congress, which provided for the transfer of federal lands to state governments, but were never reported out of committee. In addition, several states enacted statutes asserting state ownership of public lands controlled by the BLM. These efforts were rejected by the court in Nevada v. the United States.[58] President Reagan eased the pain of defeat by appointing leaders of the Rebellion to important positions and by staffing the federal bureaucracy with westerners, who managed federal lands as a good neighbor.[59]

The Wise Use Movement arose from the ashes of the Sagebrush Rebellion at the end of the Reagan administration. The movement calls for the termination of environmental controls on development, the exclusive use of federal lands by extractive industries, the protection and subsidization of commodity interests on public lands, and the expansion of motorized recreation on public lands. It claims to be a grassroots campaign, but is subsidized by the major resource development industries.[60] It supports property rights and County Supremacy and opposed Mexican wolf recovery.

The County Supremacy Movement grew out of the Wise Use Movement. From 1991 to 1994, thirty-five counties enacted ordinances asserting control over public lands despite near unanimous opposition from their state attorneys general.[61] These county ordinances followed two general models. The first protected local customs and culture associated with ranching, farming, grazing, and timber harvesting. The prime example was the 1991 Catron County New Mexico ordinance authorizing the development of a land use plan to protect local custom and culture. Federal and state governments were required to consult with and coordinate their activities with the county regarding the management of federal and state lands. Federal agencies had to submit written reports to the county regarding their activities that were subject to county approval. The federal government could not change any land use, acquire new property, designate wilderness, or wild and scenic rivers without county approval. The second model statute, which was the child of the Sagebrush Rebellion, simply declared state and county control over public lands. The "custom and culture" ordinances received a lethal blow in Boundary Packers v. Boundary County.[62] Nevertheless, these ordinances were invoked to oppose Mexican wolf reintroduction and recovery.

Conclusion

The fate of the Mexican wolf was tied to changes in public land management. The Mexican wolf was nearly driven to extinction through state bounties and federal predator control instigated by extractive users, who dominated public land policy. Scientists were very critical of these efforts.

After the Second World War, public land management changed from being exclusively focused on resource development to a more balanced approach that considered noneconomic values. The environmental movement emerged in the 1960s and 1970s. At the same time, there were fundamental changes in the political system that opened up the formerly closed congressional and administrative decision-making process to environmental groups. Standing rules were liberalized, allowing environmental groups to challenge government actions. Federal courts were willing to examine executive actions to ensure conformity with statutory and constitutional mandates.

Federal predator control became intertwined with the protection of endangered and threatened species, resulting in the enactment of the ESA. The changes in public land management generated conflicts with extractive industries and some states manifested in the Sagebrush Rebellion, the Wise Use, and Country Supremacy movements. This history provides the background for the Mexican wolf's reintroduction and recovery.

Notes

1 D. Bernard Zaleha, "The Rise and Fall of BLM's Cooperative Management Agreements," 17 Environmental Law 125, 129 (1986).

2 David E. Brown, *The Wolf in the Southwest, The Making of an Endangered Species* (University of Arizona Press, 1984), 18–19, 31–2, 41–3.

3 Robert K. Mortensen, *In the Cause of Progress, A History of the New Mexico Cattle Grower's Association* (New Mexico Stockmen, 1983), 71.
4 George C. Coggins and Parthenia B. Evans, "Predator's Rights and American Wildlife Law," 24 Arizona Law Review 821, 827–30 (1982).
5 Michael J. Robinson, *Predatory Bureaucracy, The Extermination of Wolves and the Transformation of the West* (University of Colorado Press, 2005), 28.
6 Thomas A. Lund, *American Wildlife Law* (Oxford, 1980), 60–73.
7 Brown, 48.
8 Robinson, 62–3, 381.
9 James B. Trefethen, *An American Crusade for Wildlife* (Winchester, 1975), 284. Robinson, 115.
10 Gary Lee Nunley, *The Mexican Gray Wolf in New Mexico* (Division of Animal and Wildlife Control, Department of Interior, 1977), 15.
11 Dan Miles Gish, *A Historical Look at the Mexican Gray Wolf in Early Arizona Territory and Since Statehood* (1978), 65, 74. Brown, 54.
12 Mortensen, 72.
13 Thomas R. Dunlap, "Values for Varmints: Predator Control and Environmental Ideas, 1920–1939," 53 Pacific Historical Review 141, 147–58 (1983). Coggins and Evans, 835–7.
14 Brown, 58, 63.
15 Gish, 93.
16 Brown, 63–5.
17 Id., 70, 77.
18 Mortensen, 72–3.
19 Dunlap, 147–53. Robinson, 184.
20 *Hunt v. U.S.*, 278 U.S. 96 (1928). Coggins and Evans, 832.
21 Faith McNulty, *Must They Die?* (Doubleday, 1971), 14–16, Wick Corwin, "Predator Control and the Federal Government," 51 North Dakota Law Review 787, 802 (1975).
22 Coggins and Evans, 835–7.
23 Patrick Austin Perry, "Law West of the Pecos: The Growth of the Wise Use Movement and the Challenge to Federal Public Land-Use Policy," 30 Loyola Los Angeles Law Review 275, 290–1 (1996).
24 Dunlap, 65–83. George C. Coggins, "Federal Wildlife Management," 60 Oregon Law Review 59, 64–75 (1981).
25 Coggins and Evans, 838–9.
26 Durward L. Allen, *Our Wildlife Legacy* (Funk and Wagnalls, 1962), 237.
27 Stanley P. Young and Edward A. Goldman, *The Wolves of North America* (Dover, 1944), 385.

28 Corwin, 802–3.

29 Stanley A. Cain et al., *Predator Control—1971*. Report to the President's Council of Environmental Quality and The Department of the Interior by the Advisory Committee on Predator Control (1971), 22.

30 Aldo Leopold, *A Sand County Almanac: And Sketches Here and There* (Oxford Press, 1966), 240.

31 Id., 130.

32 Theodore Lowi, *The End of Liberalism*, 2d ed. (Norton, 1969).

33 Richard Stewart, "The Reformation of American Administrative Law," 88 Harvard Law Review 167 (1975).

34 George Cameron Coggins, "Of Succotash Syndrome and Various Platitudes," 53 University of Colorado Law Review 229, 279–80 (1982).

35 Daniel J. Rohlf, *Endangered Species Act: A Guide to Its Protections and Implementation* (Stanford Environmental Law Society, 1989), 20. Coggins and Evans, 843–4.

36 James S. Bowman, "The Environmental Movement: An Assessment of Ecology Politics," 5 Environmental Affairs 649 (1976).

37 Andrew S. McFarland, *Public Interest Lobbies: Decision Making on Energy* (AEI, 1976), 1–24.

38 Robert L. Rabin, "Lawyers for Social Change: Perceptions on Public Interest Law," 28 Stanford Law Review 207 (1976).

39 Coggins and Evans, 840–3. Dunlap, 98–110.

40 Dale D. Goble, "Evolution of at Risk Species Protection," in Frank W. Davis, J. Michael Scott, and Dale D. Goble (eds.), *The ESA at Thirty* (Island Press, 2005), 59.

41 David Todd, "Wolf-Predator Control: Endangered Species Protection," 33 South Texas Law Review 459, 484–6 (1992). J. J. McCoy, *Wild Enemies* (Hawthorne Books, 1974), 157–75.

42 A. S. Leopold et al., *Predator and Rodent Control in the U.S.*, Report submitted to Stewart Udall, Secretary of Interior (March 9, 1964).

43 Corwin, 804–6. Coggins and Evans, 845–50.

44 Robinson, 306–15.

45 Public Land Law Commission, *One Third of the Nation's Land: A Report to the President and to the Congress* (Washington, DC, 1970), 168.

46 Cain et al., 2. Coggins and Evans, 851–2.

47 Corwin, 804–6. Coggins and Evans, 851.

48 37 Federal Register 2875 (1972). Coggins and Evans, 852–6.

49 37 Federal Register 5718–20 (1972).

50 June C. Edvenson, "Predator Control and Regulated Killing: A Biodiversity Analysis," 13 Journal of Environmental Law 31, 44–54 (1995).

51 Nancy E. Marion, *Making Environmental Law: The Politics of Protecting the Earth* (Praeger, 2011), 233,

52 President Nixon, "Special Message to Congress Outlining the 1972 Environmental Program" (February 8, 1972).

53 Bruce Hampton, *The Great American Wolf*, (Henry Holt & Company, 1997), 176.

54 *Alliance for the Wild Rockies v. Salazar*, 800 F. Supp. 2d 1123, 1126 (2011).

55 16 U.S.C. 1531(a) (3), (b).

56 *TVA v. Hill*, 437 U.S. 151, 177, 180–6 (1973).

57 John D. Leshy, "Unraveling the Sagebrush Rebellion: Law, Politics and Federal Lands," 14 University of California Law Review 317 (1980). Michael E. Shapiro, "Sagebrush and Seaweed Robbery: State Revenue Losses for Onshore and Offshore Federal Lands," 12 Ecology Law Quarterly 481 (1985).

58 512 F. Supp. 166, 172 (1981). A. C. Titus, "The Nevada 'Sagebrush Rebellion' Act: A Question of Constitutionality," 23 Arizona Law Review 264, 265–8 (1981).

59 Coggins and Evans, 856–62.

60 Philip D. Brick and R. McGreggor Cawley (eds.), *A Wolf in the Garden: The Land Rights Movement and the New Environmental Debate* (Rowman & Littlefield, 1996).

61 Andrea Hungerford, "Custom and Culture Ordinances: Not a Wise Move for the Wise Use Movement," 8 Tulane Environmental Law Journal 457, 460–1 (1995).

62 913 P.2d 1141, 1146–8 (Idaho, 1996). Peter D. Coppelman, "The Federal Government's Response to the County Supremacy Movement," 12 Natural Resources & Environment 30, 31–2 (1997). Robert L. Glicksman, "Fear and Loathing on the Federal Lands," 45 University of Kansas Law Review 647 (1997).

3

The Livestock Industry's Opposition to Lobo's Return

After the Mexican wolf was declared an endangered species in 1976, the federal government was required to return the Mexican wolf as a functioning component of the southwest ecosystem. Reintroduction efforts began in the Clinton administration. The Mexican wolf was reintroduced into the Blue Range Wolf Recovery Area (BRWRA) in central Arizona and New Mexico as a nonessential experimental population pursuant to section 10(j) of the Endangered Species Act (ESA). The chapter analyzes the litigation in New Mexico Cattle Growers Association v. U.S. Fish and Wildlife Service dealing with the livestock industries effort to halt Mexican wolf reintroduction during the Clinton administration and the litigation in Coalition for Arizona/New Mexico Counties for Stable Economic Growth v. U.S. Fish and Wildlife Service rejecting subsequent efforts to stop Mexican wolf reintroduction and recovery during the George W. Bush administration. Both cases focused on the ESA and National Environmental Policy Act.

3.1 The Endangered Species Act

The ESA was enacted in 1973 to address the extinction of species. Section 4 of the ESA authorizes the secretary of interior to utilize the "best scientific and commercial data" to list and delist domestic or foreign species as endangered or threatened. An endangered species is "any species which is in danger of extinction throughout all or a significant portion of its range." A threatened species is "any species which is likely to become an endangered species in the foreseeable future throughout all or significant portion of its range." Species can be listed and delisted based on the following five factors: "A) the present or threatened destruction, modification, or curtailment of its habitat or range: B) overutilization for commercial, recreational, scientific, or educational purposes: C) disease or predation; D) the inadequacy of existing regulatory mechanisms; or E) other natural or manmade factors affecting its continued existence."[1] The FWS has regulatory authority over land-based animals.

Section 7 of the ESA requires federal agencies to carry out programs for the conservation of endangered and threatened species. Each federal agency must ensure that any action that it authorizes, funds, or carries out is not likely to "jeopardize the continued existence of any listed species." Each federal agency must consult with the FWS to determine if an endangered or threatened species is present. If one is present, the agency must perform a biological assessment to determine if the species will be affected. If so determined, the agency must consult with the FWS. If, however, the agency concludes through informal consultation or the preparation of a biological assessment that the proposed action is "not likely to adversely affect" the listed species and the FWS concurs with the determination, the consultation process is ended.[2]

The ESA was amended and reauthorized in 1978 and 1979. The newly created Endangered Species Committee may provide an exemption from the stringent requirements of section 7 if the federal activity is deemed to be in the national interest. The secretary is also directed to develop and implement recovery plans that are "designed to ensure the conservation or survival of each listed species."[3] Conservation "means to use and the use of all methods and procedures which are necessary to bring any endangered species or threatened species to the point at which the measures provided pursuant to this chapter are no longer necessary." This definition found in section 4 recovery plans and section 7 provides the FWS with the authority to translocate and reintroduce endangered and threatened species back into their historic range. There is no distinction between the protection afforded to natural as opposed to reintroduced species. This has generated political opposition to reintroduction.[4]

In 1982, Congress provided a more efficient means to advance and decrease political opposition endangered and threatened species recovery. Section 10(j) of the ESA grants the FWS flexibility to establish and decrease the protection afforded to a reintroduced experimental population. Section 10(j) identifies an experimental population as "any population (including any offspring arising solely thereof) authorized by the Secretary for release under paragraph (2), but only when, and at such times as, the population is wholly separate geographically from nonexperimental populations of the same species." Section 10(j)(2) authorizes the secretary to release "any population (including eggs, propagules, or individuals) of an endangered species or threatened species outside of the current range of such species if the Secretary determines that such release will further the conservation of the species." Before authorizing such a release, "the Secretary shall by regulation identify the population and determine, on the basis of the best available information, whether or not such population is essential to the continued existence of the endangered or threatened species." All members of the experimental population "shall be treated as a threatened species." A nonessential population is only managed as a threatened species "within the National Wildlife Refuge System or the National Park System." Otherwise it is treated "as a species proposed to be listed." No critical habitat is designated for a nonessential population.[5]

3.2 The National Environmental Policy Act

The National Environmental Policy Act (NEPA) establishes a national commitment by the federal government to protect the environment and backs this commitment with action-forcing requirements. When a federal agency contemplates a major federal action that significantly affects the environment, the agency must prepare an Environmental Impact Statement (EIS), which discusses (1) the environmental impacts of the proposed action; (2) any unavoidable environmental effects; (3) alternatives to the proposed action; (4) the relationship between local short-term uses of the environment and the maintenance of long-term productivity; and (5) any irreversible and irretrievable commitment of resources. The EIS ensures that federal officials have considered environmental factors in their decision-making process. Furthermore, the EIS informs the public and other political actors about the potential consequences of the proposed federal activity.[6]

3.3 The Mexican Wolf Recovery Program

The killing of the last Mexican wolf in the United States, the Aravaipa wolf, was a "political assassination." After the Mexican wolf was declared an endangered species in 1976, Arizona ranchers, fearing that the Aravaipa wolf would become the founder of the new line of wolves, offered a bounty on the wolf, which was killed by a private trapper later in the year. Nevertheless, Mexican wolves continued to migrate into the United States from Mexico.[7]

Section 4(f) of the ESA requires the secretary of interior to develop and implement recovery plans "for the conservation and survival of endangered and threatened species." "Recovery" is the process that halts the decline of and eliminates threats to endangered and threatened species. Recovery planning first began as a voluntary agency process. In 1978, Congress amended the ESA and required recovery planning. The first Mexican wolf recovery team was formed in 1979 during the Carter administration.[8]

Ronald W. Reagan, an avowed Sagebrush rebel, won all of the western states in the 1980 and 1984 presidential elections. President Reagan's "New Federalism" promised to return many domestic issues back to the states. He criticized environmental laws and opposed federal government regulation. The Reagan administration reluctantly followed the dictates of the ESA, but did not support wolf reintroduction.

The United States and Mexico agreed to a recovery plan in 1982, which concluded there was "no possibility for the complete delisting of the Mexican wolf," but only the downlisting if it to a threatened species. The plan was designed to conserve and ensure the survival of the subspecies by maintaining a captive breeding program and reestablishing a viable, self-sustaining population of 100 wolves in a 5,000-square-mile region of its historic range.[9]

The captive breeding program was initiated with the capture of five Mexican wolves (one pregnant female, four males) by Roy McBride from 1977 to 1980 in Durango and Chihuahua, Mexico. The female gave birth to five pups (four males, one female). The female pup died soon after birth. In 1980, McBride captured a male Mexican wolf near Chihuahua, which bred with the initial surviving female, producing a litter in 1981. Three of the animals (two males, one female) produced offspring that established the certified captive lineage, also known as the McBride lineage. In 1981, Mexico informed the recovery team that only thirty wolves remained in Mexico. Mexico joined the captive breeding program in 1987. By 1994, the binational breeding program produced a captive breeding population of ninety-two wolves.

Two additional populations determined to be purebred Mexican wolves were certified in July 1995. The US population known as the Ghost Ranch line consisted of twenty-one known animals, all descended from two founders. The original sire was captured in 1959 near Tumacacori, Arizona. The founding female was purchased in 1961 by a tourist in Mexico, so it was not known if she was born in the wild or in captivity.

The Mexican population known as the Aragon line consisted of eight animals in the San Juan de Aragon Zoo in Mexico City. Their origin was unknown. All three lines were pure Mexican wolves. Two of the four members of the certified lineage were probably mother and son, reducing the unrelated members to three. The total captive population in the United States is descended from the seven founders.[10]

The recovery plan developed during the Reagan administration, just months before the enactment of section 10(j), laid dormant because of a lack of funding, minimal administrative commitment, and political opposition. The FWS decided not to proceed without the support of state fish and game commissions. This almost doomed the program because the Arizona, New Mexico, and Texas Game and Fish Commissions, which develop state wildlife policy, opposed Mexican wolf reintroduction.

National and state environmental groups pushed for reintroduction in 1986. The FWS, echoing the state rights claims of the Reagan administration, asked the states to identify proposed sites. Texas stonewalled. Arizona identified fifteen sites, none of which had been evaluated. New Mexico suggested White Sands Missile Range, where 4,000 square acres provided enough room for forty wolves. The army, following the FWS policy, initially refused to allow the release of Mexican wolves without state approval.[11]

The 1988 ESA amendments added additional requirements for recovery plans. Recovery plans had to contain site-specific management actions, objective measurable criteria for delisting, and estimates of the time and costs to restore the species. The amendments also provided the opportunity for public comment on new or revised recovery plans.

George H. W. Bush won most of the western states, including Arizona and New Mexico, in his 1988 presidential victory. President Bush promised to be the environmental president. He acknowledged the importance of wolf reintroduction but took no steps to realize wolf recovery. In April 1990, a coalition of environmental groups sued the Department of Interior and the US Army for not implementing the wolf recovery plan. They alleged that the FWS violated section 4(f)(1) of the ESA by failing to establish a wild population of Mexican wolves. The Defense Department violated section 7(a)(1) by failing to "carry[ing] out programs for consideration of endangered species," particularly by the withdrawal of White Sands as a potential release site in 1987. In addition, they argued it was not appropriate to give the state and local governments veto power over federal action.[12]

Shortly after the notice of intent to sue was filed, the army reversed its decision regarding White Sands. In October 1990, David Parsons was appointed the full-time Mexican wolf recovery coordinator with instructions to expedite the recovery plan. In 1991, the FWS developed a general plan for Mexican wolf reintroduction. The timing was inopportune because of the federal deficit and competition with the red and gray wolf reintroductions.

William J. Clinton took office in 1992. President Clinton won New Mexico in 1992 and New Mexico and Arizona in 1996. Former Arizona governor Bruce Babbitt, a strong supporter of wolf recovery, became secretary of interior. The Clinton administration focused on returning gray wolves to the Northern Rocky Mountains but was also concerned with Mexican wolf reintroduction.

The Clinton administration negotiated a settlement agreement in the 1990 suit. Settlement agreements have certain advantages and disadvantages. The parties themselves negotiate the terms of the settlement, providing certainty to the outcome. Settlement is quicker and less costly than litigation. However, settlement agreements do not have any precedential value and are not subject to appeal. The FWS agreed to implement the recovery plan as expeditiously as possible and release the Draft EIS (DEIS) and the proposed rule for reintroduction by May 1994. The Final EIS (FEIS) and final rule would be issued by March 1995. The release of the Mexican wolves was scheduled for July 1996. Funding for Mexican wolf recovery had been stalled at $150,000/year. With the aid of Senator DeConcini (D. N.M.) and Representative Sidney Yates (D. Ill.), the chair of Interior Appropriations subcommittee, funding was increased to $400,000 in 1994. This allowed the EIS process to move forward.[13]

The DEIS was not released until 1995, fourteen months after the date established by the settlement agreement. The Apache and Gila National Forests and White Sands were identified as potential reintroduction sites. A survey by the League of Women Voters indicated that 62 percent of the New Mexico respondents supported, while 22 percent opposed, wolf reintroduction. In the four affected rural counties, 50 percent supported and 30 percent opposed wolf reintroduction.[14]

Despite public support, the plan faced an uphill fight. A bill was introduced in the Arizona legislature that provided a bounty for the killing of endangered wolves. Senator Domenici (R. N.M.), who was strongly supported by the livestock industry, questioned reintroduction. The New Mexico Cattle Growers Association (NMCGA) declared that the plan would evict rural residents, destroy their culture, and override property rights. The issue was not the wolf, but federal intrusion. The New Mexico Wool Growers Association (NMWGA) complained of federal government oppression in the form of inadequate predator control, rangeland reforms, and excessive federal regulations. The New Mexico Farm and Livestock Association viewed wolf reintroduction as a threat to small business and as a misuse of the ESA.[15]

Governors Gary Johnson of New Mexico and Stuart Symington of Arizona opposed the reintroduction citing the fear of rabies, negative economic impacts, bad science, and the potential change in status to an essential population. The New Mexico Game and Fish Commission (NMGFC) opposed release on the grounds that there were no suitable potential Mexican wolf release sites in the state. The Arizona Game and Fish Commission (AZGFC) only supported wolf reintroduction into White Sands.[16]

Facing impending litigation, the FWS released the FEIS in December 1996, twenty-one months later than the scheduled March 1995 release date. This precluded any possibility of wolves being released in the spring of 1997. In March 1997, the secretaries of interior and agriculture, along with the deputy secretary of the army, agreed to allow the reintroduction to proceed. In July 1997, the AZGFC agreed to cooperate with wolf recovery.

The final rule for the reintroduction of Mexican wolf into the BRWRA as a nonessential experimental population pursuant to section 10(j) of the ESA was published in January 1998.[17] The BRWRA, which covered 6,854 square miles, consisted almost entirely of national forest lands, including the Apache and Gila National Forests. The BRWRA was within the historic range of the Mexican wolf and separate from any known naturally occurring population of wild wolves.

The BRWRA was located within southern Apache and northern Greenlee counties in Arizona, and southern Catron, northern Grant, and western Sierra Counties in New Mexico. Approximately two-thirds of Greenlee and Catron counties were in the BRWRA. Apache, Catron, Greenlee, and Sierra Counties, all supporters of County Supremacy, had land use ordinances that established the environmental planning and review processes and required federal decisions within the counties to be subject to local approval. Catron and Sierra Counties also enacted ordinances prohibiting the release of wolves into the wild. Each of the affected counties opposed the reintroduction of the Mexican wolf and asserted that the reintroduction would threaten local custom and culture. The FEIS declared that the reintroduction would not violate the Catron and Sierra

County ordinances prohibiting the release of wolves, but warned that federal law would preempt any county ordinances.

The designation of the Mexican wolf as a nonessential experimental population provided the secretary with flexibility. A reintroduced Mexican wolf in a national park or wildlife refuge was considered a threatened species, so formal consultation with the FWS was required. A reintroduced Mexican wolf located outside of a national park or wildlife refuge was treated as a species proposed for listing, so the federal agency had to only confer with the FWS regarding any actions likely to jeopardize the species. The nonessential experimental population designation allowed for harassment and limited taking of the wolf, if the wolf threatened a human or was killing livestock.[18]

The FWS planned to move Mexican wolves to the Sevilleta National Wildlife Refuge in central New Mexico, where they would be paired and moved into release pens. Initially, a soft release was planned. The wolves would be held in pens for several weeks for acclimation. Fourteen groups would be released over a five-year period, with the goal of reaching 100 wolves in the BRWRA. Initially, the wolves would only be released into the Apache National Forest in Arizona and allowed to disperse into the Gila National Forest in New Mexico. Problem wolves would be captured and relocated into the Gila National Forest. The wolves would not be allowed to establish territory beyond the BRWRA.[19]

In January 1998, eleven wolves in three family groups were brought to pens in preparation for release. Secretary Babbitt welcomed the wolves in the face of rancher protest. Several livestock groups brought suit against the FWS to halt the release, alleging NEPA and ESA violations. The wolves were quietly released into eastern Arizona.[20]

Arizona resident Richard Humphrey killed the first Mexican wolf after it allegedly threatened his dog. He later changed his story, claiming self-defense. The FWS, hoping to bank some goodwill, refused to prosecute. This angered wolf supporters, who predicted an open season on wolves.[21] In 1998, five of the thirteen wolves released were shot. One was missing and presumed dead. Three were recaptured. One pup born in the wild disappeared after its mother was shot. Only three males remained in the wild. An underground group, the Minutemen, claimed credit for the killings. A $50,000 reward was offered for the apprehension of the killers.[22]

In September 1999, David Parson, director of the Mexican wolf project, proposed the translocation of problem wolves into New Mexico. Translocation into the Gila National Forest had been discussed in general terms in the FEIS.

Translocation could limit wolf–human conflicts, halt wolf losses, and aid in wolf dispersal into suitable habitat. The Gila National Forest area contained 700,000 acres where there were no federal grazing permits. David Parsons retired later in September. He was supposed to be rehired, but the FWS reneged on the promise. Program advocates questioned why Parsons was not rehired.[23]

The proposed translocation into New Mexico was opposed by Catron and Grant Counties. Representative Joe Skeen (R.N.M.), an opponent of wolf reintroduction, demanded that the release be delayed, claiming the EIS was outdated. Representative Skeen, a member of the House Appropriations Subcommittee on Interior, threatened the funding for the project.[24] In 1999, twenty-two wolves were released, including ten adults and twelve pups born in acclimation pens.

3.4 New Mexico Cattle Growers Association v. U.S. Fish and Wildlife Service

The NMCGA brought suit against the FWS, challenging the Mexican wolf reintroduction program on NEPA and ESA grounds. Other livestock groups[25] and environmental groups[26] intervened in the suit. Intervention is a process that allows a party, who is not an original party in the suit, but who will be directly and substantially be affected by the outcome, to participate in the case as a party. In October 1999, the U.S. District Court for the District of New Mexico accepted the FWS analysis and rejected the NMCGA contentions regarding livestock depredation rates, the hybridization of the reintroduced population, the existence of a naturally occurring population of Mexican wolves in the BRWRA, the impacts on other endangered and threatened species, federal consultation with state and local governments, and the need for a Supplemental EIS (SEIS).[27]

Judicial review of the FWS compliance with the NEPA and ESA is governed by the Administrative Procedures Act (APA), which is subject to the hard-look doctrine. The court must determine if the agency's action is "arbitrary, capricious, an abuse of discretion, or otherwise not according to the law" or "without observance of the procedure required by law."[28] The court must ensure that the agency has considered all relevant factors and has made a reasonable judgment based on these factors. The court's inquiry must be "searching and careful," but the scope of review is limited. The court cannot substitute its judgment for that of the agency. Even though the agency's decision is entitled to a "presumption of regularity," the presumption does not protect the agency's action from a thorough, probing, in depth review."[29]

3.4.1 Livestock Depredation

The FWS examined depredation rates in Minnesota, Montana, and Alberta and utilized the Yellowstone reintroduction equation to estimate wolf depredation in the BRWRA. The FWS estimated that one to thirty-four head of cattle would be killed by each wolf, which constitutes 0.001 percent to 0.04 percent of the 82,600 cattle in the BRWRA.[30]

The NMCGA argued that the FWS methodology was flawed and failed to consider relevant contrary information. The FWS failed to consider the ratio of wild to domestic prey in the BRWRA, so did not utilize the best available science. Two studies of wolf depredation in regions in Italy and Spain, which were similar to the BRWRA, indicated that the wolf is a nonselective predator, which consumes wild and domestic prey in proportion to its availability. Wolves do not select wild over domestic prey, but prefer smaller and younger animals. Both of these studies were published in scientific journals prior to the publication of the FEIS, but were not cited by the FWS. In addition, the FWS failed to consider the killings by offspring of the 100 reintroduced wolves and unconfirmed livestock losses from wolf depredation. A later more accurate study concluded that 500 to 700 cattle and 120 to 150 sheep would be taken each year by Mexican wolves.[31]

The court refused to consider the excluded studies because they were not part of the administrative record. The court concluded that the FWS had adequately analyzed wolf depredation. The NEPA only required the agency to adequately consider the issue and make a reasonable decision. The FWS did not have to consider every study. There was no showing that the FWS selectively excluded the studies, two of which had not been completed prior to the FEIS. The FWS methodology was appropriate and entitled to substantial judicial deference. Disagreement among experts was not a basis for invalidating the EIS.[32]

3.4.2 Hybrid Wolves

The FWS employed a six-person committee, chaired by Dr. Phillip Hedrick, to examine the genetic lines of the reintroduced population. The committee considered the morphology and molecular genetics of the breeding lines, as well as molecular techniques, including mitochondrial DNA, multi-locus DNA fingerprinting, and microsatellite loci analysis. The committee concluded that the captive wolves were not hybrids, but pure Mexican wolves.[33]

The NMCGA alleged that the FEIS failed to consider the possibility that the reintroduced wolves were not genetically pure, but had crossbred with dogs and coyotes. The ESA did not authorize the release of hybrids, which would not preserve the species. The Ghost Ranch line, whose founder's lineage could not be traced back, manifested dog characteristics. Testimony by Roy McBride, the original trapper, and Curtis Carley, an original member of the recovery team, supported this position. The FWS determination of genetic purity of the Ghost Ranch line was dubious. Studies of the wolves in Minnesota showed that 50 percent of the wolves were hybrids.[34]

The court found that even though the FEIS did not consider the hybrid issue, the FWS had thoroughly analyzed the issue. Expert opinion showed the reintroduced wolves came from three breeding lines free from introgression by dogs and coyotes and were "genetically pure" Mexican wolves.[35]

The court concluded that the evidence supported the FWS's conclusions regarding the hybridization issue. The FWS genetic committee utilizing state-of-the-art genetic techniques determined that the three founding lines were Mexican wolves. The findings were confirmed by another study comparing the genetic data from the Ghost Ranch and Aragon lines performed by scientists at UCLA and the Institute for Nature Conservation Research at Tel Aviv University. The scientists compared the "microsatellite polymorphisms" of 151 gray wolves, 142 coyotes, 42 dogs. The studies concluded that the Mexican wolves "do not have alleles specific to domestic dogs or coyotes, but rather share alleles found in certified Mexican wolves." The two lines of captive wolves "are unlikely to have been founded by a simple cross between grey wolf and domestic dog or coyote but likely represent the descendants of pure Mexican wolves." Since the wolves are genetically pure, they "should be used as a source for reintroduction to the southwestern United States, if possible."[36]

The NMCGA relied on the work of Dennis Parker, who wrote an unpublished paper in 1987, "Southwest Wolves: Discussion of Their Taxonomical Arrangement." The paper was revised in 1995, "Reintroduction of the Mexican Wolf: Instrument of Recovery or Instrument of Demise?" and served as a comment on the DEIS. The FWS put together a six-person genetic committee of eminent scientists to examine Parker's assertions. The results appeared in an appendix to the FEIS. The committee was very critical of Parker's work. David Mech, a noted wolf expert, declared that the paper "misinterprets and misconstrues" the evidence and "ignores other salient studies." The paper was not "an objective analysis," but an effort to prevent the reintroduction of wolves.[37]

The evidence submitted by the NMCGA was dubious. Roy McBride was an experienced trapper, not an expert on molecular genetics. Curtiss Carley was not a genetic expert, but relied on earlier techniques that were not as reliable as current techniques. Furthermore, the hybridization of wolves in Minnesota was not relevant to the Mexican wolf.[38]

3.4.3 Presence of a Naturally Occurring Wolf Population

The FWS determined that there was no evidence that Mexican wolves were present in the BRWRA. The FWS analyzed all sightings of wolves in Arizona, New Mexico, Texas, and northern Mexico from 1983 to 1993. The FWS interviewed individuals and conducted field surveys regarding sightings after 1993. Technicians solicited howling responses at 1,727 howling stations along a 1,600-mile route. Searches for wolf tracks and scat were also conducted.[39]

The NMCGA argued that there was a native population of Mexican wolves in the BRWRA. There were sightings of individual wolves, which constituted an existing population. Section 10(j) prohibited the reintroduction of a nonessential experimental population in the current range of the species.[40]

The court found no support for NMCGA's conclusion except anecdotal information. The FEIS acknowledged that since 1983 there had been six unconfirmed reports about wolf sightings in the BRWRA. The last verified sighting of the Mexican wolf in New Mexico occurred in 1976 and in Arizona in 1983 or 1995. The FWS conducted wolf howling surveys in seven different habitat areas. The FWS also did scent monitoring, surveys of howl routes, searches for dens and killing sites, hunts for tracks, and track castings. The FWS employed recording devices and remote cameras, and conducted formal investigations of wolf sightings. The evidence demonstrated that no Mexican gray wolves were present in the BRWRA.[41]

The Ninth and Tenth Circuits had previously addressed this issue. Both courts concluded that anecdotal sightings of lone dispersers in the wolf recovery area did not constitute a population.[42] In addition, the FEIS noted that if there was a wild Mexican wolf in the area, it could interbreed with the reintroduced wolves and enhance genetic diversity of the population.[43]

3.4.4 Consideration of Other Endangered Species

Section 7(a) (2) of the ESA requires each federal agency to ensure its action "is not likely to jeopardize the continued existence of any listed species." After informal consultation with the U.S. Forest Service and intra-agency consultation, the FWS concluded that implementation of the Mexican wolf reintroduction program would not have any adverse effect on any listed species.[44]

The NMCGA alleged that the FEIS failed to consider the impacts on endangered and threatened species. The FEIS listed fifteen endangered and threatened species and another sixty-one scheduled for listing, but only contained a limited discussion of the impacts of Mexican wolf reintroduction. The FWS did not conduct any analysis because the wolf would not prey on these species and later review would occur. There were no studies about indirect impacts of wolf reintroduction on vegetation and other animals in the BRWRA. For example, the FEIS noted that the Mexican spotted owl and Mexican wolves might overlap, but no data was presented that the wolf would not jeopardize the owl. The failure to analyze impacts on endangered and threatened species violated the NEPA and ESA.[45]

The court held that the FEIS adequately considered the impact of the recovery program on endangered and threatened species in the BRWRA. The FWS collected and considered data on numerous species in Arizona and New Mexico, particularly endangered and threatened species. The court determined that experts disagreed on the impacts and little consensus existed regarding the management of endangered and threatened species. The needs of the species were often in opposition to one another. Just because the FWS did not resolve the conflict did not mean it did not consider all relevant factors and make a reasonable decision. The FEIS included all relevant and essential information for making an informed decision.[46]

The NMCGA complained about the summary of the information on endangered species in the FEIS. However, this was not the only data considered by the FWS. For example, the threat to the Mexican spotted owl was considered in another document, the Mexican spotted owl recovery plan. The FWS determined that the Mexican wolf might, but was not likely, to impact the Mexican spotted owl because their principal prey did not generally overlap. The Mexican wolf took large ungulates and only went after small mammals, the prey of the Mexican spotted owl, when large ungulates were unavailable. The FWS noted that the wolf might help owl recovery by reducing overgrazing by deer and elk in the BRWRA.[47]

3.4.5 Federal Consultation

The federal regulations regarding experimental populations require the FWS to "consult with appropriate State fish and wildlife agencies, local governmental entities, affected Federal agencies, and affected private landowners in developing and implementing experimental population rules." Any regulation under this section "shall, to the maximum extent practicable, represent an agreement between the FWS, the affected State and Federal agencies and persons holding any interest in land which may be affected by the establishment of an experimental population."[48]

The NMCGA alleged that the FWS failed to comply with the public comment and consultation requirement. The FWS ignored the complaints of landowners regarding the destruction of their livestock and private property. The FWS was required to enter into private agreements with affected private landowners. Furthermore, the FWS failed to discuss coordination with state and local governments, particularly regarding the counties' custom and culture ordinances.[49]

The court, upholding the FWS determination, found the FEIS contained a whole section on consultation and coordination. There were four public hearings prerelease and extensive public input from the Arizona and New Mexico Game and Fish Departments, the White Mountain and San Carlos Apache tribes, and the Departments of Agriculture and Defense. The FWS kept the local public informed and worked with local officials to generate support for the proposal. Nevertheless, it was very difficult to get support for the program. The FWS did not neglect or deal with this issue in an arbitrary and capricious manner.[50] The court did not address the counties' custom and culture ordinances.

The court's decision was supported by the text of the regulation and legislative history. The text of the regulation only required a good faith effort to achieve such an agreement prior to reintroduction. In addition, the legislative history indicates Congress only intended that "such rules and regulations, to the maximum extent practicable, serve as a type of cooperative agreement, between the affected parties." There was no requirement that the FWS reach agreements with all affected government agencies and landowners.[51]

3.4.6 Supplemental Environmental Impact Statement

The NMCGA alleged that a Supplementary Environmental Impact Statement (SEIS) was necessary. The FEIS failed to disclose all relevant information. The Spanish and Italian studies on wolf depredation were not part of the administrative record and needed to be circulated and commented upon. The FWS comment on the studies did not remedy the defect, but was just a post hoc rationalization. Furthermore, the new sightings of Mexican wolves in the BRWRA contradicted the FWS assumption that no natural wolves were present in the BRWRA.[52]

The court upheld the FWS determination. A SEIS only needed to be filed when there were "significant new circumstances or information relevant to environmental concerns bearing on the proposed action or its impacts." What was "significant" was left to the agency's discretion. The court could only override the agency's decision if it was arbitrary and capricious. Nothing showed that the FWS failure to conduct a SEIS was arbitrary and capricious.[53]

3.5 The George W. Bush Administration

The Bush administration came to office in 2001. The first three-year review of the Mexican wolf program was completed in June 2001. The Mexican wolf population had increased from four in 1998 to twenty-six by the end of 2001. The White Mountain Apache Tribe, unlike the San Carlos Apache Tribe, allowed the release of Mexican wolves on their reservation. The FWS rated the success of the program as "an 8 out of 10."[54] However, opposition to Mexican wolf recovery continued. Several relocated wolves killed livestock in Catron County. Additional Mexican wolves were relocated into Gila National Forest in New Mexico. The first Mexican wolf was killed by the federal government for cattle depredation. Thirteen Mexican wolves were shot and run over in Arizona and New Mexico. Nevertheless, the Mexican wolf population grew to fifty-five in 2003.

3.6 Coalition for Arizona/New Mexico Counties for Stable Economic Growth v. U.S. Fish and Wildlife Service

The livestock industry rekindled the earlier litigation in April 2003, opposing Mexican wolf recovery. The Coalition for Arizona/New Mexico Counties for

Stable Economic Growth (CSEG), which consisted of livestock and agrarian interests,[55] brought suit, seeking an injunction to halt any further releases and recapture the Mexican wolves in the BRWRA. To issue an injunction, the federal district court had to assess the likelihood of the plaintiffs' success on the merits, whether the plaintiffs would suffer irreparable harm during the pendency of the litigation, whether the balance of equities favored the plaintiff, and whether the injunction was in the public interest.[56] The Coalition alleged that the discovery of hybrid wolves in the Pipestem pack required intra-agency consultation and violated the conservation mandate in section 10(j) of the ESA. Hybridization would permit the introgression of canine genetic material into the wolf population. Continual breeding by and with hybrids would lead to the extinction of pure genetic Mexican wolves. The Coalition also contended that a SEIS was necessary to discuss the translocation of problem wolves into the secondary recovery area; the resulting depredation rates in the secondary area; and the hybridization of Mexican wolves.[57]

The FWS responded that intra-agency consultation with noted experts demonstrated that hybridization in the wild was rare and not a significant threat. The breeding cycle of wolves and dogs was different. Steps taken by the FWS would prevent hybridization. The best way to avoid hybridization was to increase the population of wild Mexican wolves. The FWS argued that a SEIS was not necessary because all of the issues had been adequately discussed in the FEIS. Furthermore, any cessation of the relocation or the removal from the wild would jeopardize Mexican wolf recovery.[58]

In July 2004, the U.S. District Court for the District of New Mexico denied the livestock industry request for a preliminary injunction in CSEG v. FWS. The court held that the reintroduction and translocation of Mexican gray wolves should continue because it would "further the conservation of the species and thereby advance the congressional priorities set forth" in the ESA.[59]

The case on the merits was argued in U.S. District Court in the District of New Mexico, which found the plaintiffs' claims baseless.[60] The court pointed out that the issues in the case required deference to administrative expertise. Such deference was consistent with the Supreme Court's recent admonishment that section 706 of the APA was designed to protect agencies from judicial interference with their lawful discretion, and to avoid judicial entanglement in abstract policy disagreements which courts both lack the expertise and information to resolve.[61] The New Mexico Farm Labor Bureau vowed to continue its opposition, declaring: "We want this program stopped. We're the ones suffering from it, and we're not gonna give up. There are other remedies available in court, and in the court of public opinion."[62]

Conclusion

The livestock industry, one of the principal opponents of reintroduction, brought suit challenging Mexican wolf reintroduction in the Clinton administration. The federal district court in NMCGA v. FWS rejected the livestock industry's assertions regarding wolf depredation, hybrid wolves, naturally occurring wolves, and other endangered species and upheld the FWS decision to reintroduce the wolf. The livestock industry continued to oppose Mexican wolf reintroduction during the Bush administration, alleging that excessive depredation, hybridization, and translocation undermined the program. The federal district court in CSEG v. FWS continued to reject livestock industry assertions and upheld the FWS decisions, allowing Mexican wolf recovery to proceed.

Both judicial skeptics and judicial proponents would approve of these decisions. Judicial skeptics, who are critical of judicial interference with the federal implementation of policy, would support these decisions because the court adopted a deferential position upholding the FWS actions. Judicial proponents, who favor judicial oversight of administrative decisions to ensure statutory compliance, would also support these decisions. The federal district courts upheld the FWS decisions, but only after ensuring that the FWS took a "hard look" at all the relevant factors.

The reintroduction of the Mexican wolf was an important victory for environmental groups against resource developers in the war for the West. Cass Sunstein points out that the law not only prescribes behavior, but also expresses underlying cultural values.[63] Robert Keiter aptly observed that the "restoration of the wolf can be viewed as an end to an era of agricultural dominance on the public domain and the ascendancy of ecosystem-oriented resource management policies."[64]

Notes

1 16 U.S.C. 1532 (6), (20). 1533(a).

2 16 U.S.C. 1536 (a) and (b). 50 C.F.R. 402.13, 402.14 (a), (b).

3 16 U.S.C. 1540, 1533 (f) (2001).

4 16 U.S.C. 1532. Mimi S. Wolok, "Experimenting with Experimental Populations," 26 Environmental Law 10018, 10019 (January 1996).

5 16 U.S.C. 1539 (j)(2)(C). Holly Doremus, "Restoring Endangered Species: The Importance of Being Wild," 23 Harvard Environmental Law Review 1, 18–31 (1999).

6 42 U.S.C 4332 (C). Edward A. Fitzgerald, "The Rise and Fall of Worst-Case Analysis," 18 University of Dayton Law Review 1, 4–8 (1992).

7 41 Fed. Reg. 17742 (April 28, 1976). 43 Fed. Reg. 9607 (March 9, 1978). D. R. Parsons, "Case Study: The Mexican Wolf," 36 New Mexico Journal of Science 101–23 (1996) (Parsons).

8 16 U.S.C. 1533. Federico Cheever, "Recovery Planning, the Courts and the Endangered Species Act," 16 Natural Resources & Environment 106 (2001).

9 FWS, *Mexican Wolf Recovery Plan 1982*, 23–4.

10 Parsons, 110–14.

11 Steve Grooms, *The Return of the Wolf* (Cowles, 1993), 155–60. Michael Robinson, *Predatory Bureaucracy* (University of Colorado Press, 2005), 350, 353.

12 Parsons, 117–18. *Wolf Action Group et al. v. United States*, Civil Action No. CIV-90-0390-HB, U.S. Dist. Crt. New Mexico.

13 Bobbie Holaday, *Return of the Mexican Wolf* (University of Arizona, 2003), 50–64, 70–6.

14 Defenders of the Wildlife, "Restoring the Mexican Wolf, Mexican Wolf Chronology" (www.defenders.org).

15 Keith Easthouse, "Recovery Plan Faces Uphill Battle," Santa Fe New Mexican (June 28, 1995).

16 Holaday, 100. Tony Davis, "A Controversial Creature," Albuquerque Tribune (November 15, 1995).

17 63 Fed. Reg. 1752 (January 12, 1998).

18 Id. at 1754–6.

19 U.S. Fish and Wildlife Service, "Final Environmental Impact Statement-Reintroduction of the Mexican Wolf Within Its Historic Range in the Southwestern United States" (1996) (FEIS), 3–10, 4–11, 12, 5-58-74.

20 Mike Taugher, "Ranchers Sue to Block Wolf Release," Albuquerque Journal (March 28, 1998) John Hill, "Wolves Quietly Released near N.M.," Albuquerque Journal (March 31, 1998).

21 Editorial: Hunt Down Wolf Killers, Arizona Daily Star (November 12, 1998).

22 Tim Molloy, "Babbitt Releases Wolf, Sends Message to Reintroduction Foes," Associated Press State & Local Wire (November 17, 1998). "Underground Group Takes Responsibility for Shooting," Albuquerque Tribune (December 12, 1998).

23 Wes R. Smalling, "Wild Wolves and Red Tape," Santa Fe New Mexican (December 2, 1999).

24 Mike Taugher, "Skeen Opposes Gila Wolf Project," Albuquerque Journal (December 23, 1999), Mike Taugher, "Grant County 'Prohibits' Wolf Release," Albuquerque Journal (January 14, 2000).

25 Intervenors included New Mexico Public Lands Council, Livestock Bureau, New Mexico Wool Growers Association, New Mexico Farm & Livestock Bureau, Hidalgo County Cattle Growers Association, Greenlee County Cattle Growers Association, Production Credit Association of New Mexico, Coalition of Arizona/New Mexico Counties for Economic Growth.

26 The defendant-intervenors were Defenders of Wildlife, National Parks and Conservation Association, Preserve Arizona's Wolves, Southwest Center for Biological Diversity, White Sands Wolf Coalition, Sky Island Alliance.

27 *NMCGA v. U.S. Fish and Wildlife Service*, Civ. No. 98-367M/JHG (October 28, 1999) (NMCGA).

28 5 U.S.C. 206(2)(A), (D).

29 *Citizens to Preserve Overton Park v. Volpe*, 401 U.S. 402, 416 (1971).

30 Federal Defendants Response Brief (Defendants Brief), 41–6.

31 Plaintiffs Opening Brief on the Merits (Plaintiffs Brief), 15–23.

32 NMCGA, 48–64.

33 Defendants Brief, 20–4.

34 Plaintiffs Brief, 27–31.

35 NMCGA, 64–7.

36 Defendants Brief, 21–3.

37 FEIS, Appendix K, "Response to Mr. David Parker's Comments on the DEIS," K-1-9. Holaday, 90–4.

38 Defendants Brief, 23–4.

39 Defendants Brief, 24–8.

40 Plaintiffs Brief, 29–31.

41 NMCGA, 67–71.

42 *U.S. v. McKittrick*, 142 F.3d 1170, 1174–5 (Ninth Circuit, 1998). *Wyoming Farm Bureau v. Babbitt*, 987 F. Supp. 1349 (D. Wyo. 1997); 199 F.3d 1224 (10th Cir. 2000).

43 FEIS, 5–59.

44 Defendants Brief, 28–9.

45 Plaintiffs Brief, 31–2.

46 NMCGA, 70–5.

47 Defendants Brief, 28–32.

48 50 C.F.R. 17.81 (d).

49 Plaintiffs Brief, 23–6.

50 NMCGA, 75–9.
51 987 F. Supp., 1365–6.
52 Plaintiffs Reply to Defendants Responses to Plaintiffs Opening Brief on the Merits, 17–21.
53 NMCGA, 78–81.
54 H.R. Rep. No. 107–564, at 363 (2002).
55 Five of the nine organizations were plaintiffs in the earlier suit: New Mexico Cattlegrowers Association, Grant County Farm and Livestock Bureau, New Mexico Farm and Livestock Bureau, New Mexico Public Lands Council, New Mexico Wool Growers. The other three plaintiffs were the Gila Permittees Association, Mimbres Farm and Livestock Bureau, and Arizona Cattle Growers Association.
56 *Prairie Band of Potawatomi Indians v. Pierce*, 253 F.3d 1234, 1246 (10th Cir. 2001).
57 Memorandum in Support of a Motion for a Preliminary Injunction, 12–33.
58 Defendant-Intervenors Response in Opposition to Plaintiffs Motion for a Preliminary Injunction, 10–11, 14–31.
59 *Coalition of Arizona/New Mexico Counties for Stable Economic Growth v. U.S.* FWS No. CIV 03-508 MCA/LCS, 30–51 (D.N.M. July 6, 2004).
60 *Coalition of Arizona/New Mexico Counties for Stable Economic Growth v. U.S.* FWS No. CIV 03-508 MCA/LCS, 58–62 (D.N.M January 31, 2005).
61 Id. at 27, citing *Southern Utah Wilderness Alliance v. Norton*, 124 S. Ct. 2373, 2381 (2004).
62 "Wolves: Judge Dismisses Legal Challenges to Southwest Reintroduction Program," Land Letter (February 10, 2005).
63 Cass R. Sunstein, "On the Expressive Function of Law," 144 University of Pennsylvania Law Review 2021, 2025–9 (1996).
64 Robert B. Keiter and Patrick T. Holscher, "Wolf Recovery Under the ESA: A Study in Comparative Federalism," 11 Public Land & Resources Law Review 19, 34 (1990).

4

State Interference with Mexican Wolf Recovery

Mexican gray wolf recovery generated conflict between the federal and state governments. Traditionally, the states owned and managed wildlife within their borders. English common law recognized wildlife management as a royal prerogative. This authority passed to the colonies, then to the states upon independence. The states hold wildlife in trust for the public. However, the federal government has the constitutional authority to preempt state law. The federal government has relied on various constitutional provisions, such as the treaty clause, the property clause, and the commerce clause to assert control over threatened and endangered species under the Endangered Species Act (ESA). Nevertheless, the ESA encourages cooperative federal-state management agreements, which are accompanied by the provision of federal funds.

The George W. Bush and Barack H. Obama administrations experienced problems with New Mexico and Arizona regarding Mexican wolf recovery. This chapter analyzes the settlement agreement in Defenders of Wildlife v. Tuggle, which terminated the Adaptive Management Oversight Committee (AMOC); the litigation in Wild Earth Guardians v. Lane permitting trapping in the Blue Range Wolf Recovery Area (BRWRA); and the litigation in New Mexico Game and Fish Department v. U.S. Fish and Wildlife Service overriding the state's refusal to allow the release of Mexican wolves into the state until a new recovery plan was in place.

4.1 The George W. Bush Administration

President Bush came to office in 2001. He carried Arizona in the 2000 and both Arizona and New Mexico in the 2004 presidential elections. The Bush administration criticized the pro-environmental policies of the Clinton administration and promised to consult with western states regarding public land policies. The administration, following President Reagan's New Federalism, delegated Mexican wolf management authority to the affected states. In addition, the administration began the construction of border barriers that impaired Mexican wolf recovery.

New Mexico's democratic governor William Richardson took office in 2003. Governor Richardson, unlike his predecessor republican Gary Johnson, supported Mexican wolf recovery. New Mexico joined the 2003 Memorandum of Understanding (MOU), which defined and formalized the role of each participant in the recovery program. The 2003 MOU established the AMOC, which provided guidance to the interagency field team on policy and coordinated agency activities. The AMOC's membership included the Arizona Game and Fish Department (AZGFD), New Mexico Game and Fish Department (NMGFD), U.S. Fish and Wildlife Service (FWS), the Animal Wildlife Service (WLS), the U.S. Forest Service (FS), the White Mountain Apache Tribe, Graham, Greenlee, and Navajo Counties in Arizona, Catron and Sierra Counties in New Mexico, and the New Mexico Department of Agriculture. The AMOC, which was chaired by the AZGFD, was delegated primary authority over Mexican wolf recovery. State agencies became managers, rather than advisers. The FWS was reduced to an advisory role.[1]

Before the establishment of the AMOC, the Mexican wolf population had grown steadily from four in 1998, to fifteen in 1999, twenty-two in 2000, twenty-six in 2001, forty-two in 2002, and fifty-five in 2003, meeting earlier defined population objectives. In 2005, the AMOC approved Standard Operation Procedure 13 (SOP 13), which required the permanent removal of any Mexican wolf that engaged in three livestock depredations in a one-year period. This jeopardized recovery efforts because the genetic value of the removed wolves, their importance to the pack, the time of year related to breeding and depredation incidents, and livestock husbandry practices were not considered. Removals under SOP 13 increased. Prior to SOP 13, two wolves were removed in 2003 for livestock depredation and only one in 2004. After SOP 13, six wolves were removed for depredation in 2005, fifteen in 2006, and nineteen in 2007. However, only ten cows per year had been taken by wolves in the BRWRA through October 2007.[2]

Catron County, New Mexico, the center of the County Supremacy Movement, vigorously opposed Mexican wolf reintroduction. The Catron County Commission (CCC) in February 2007 passed an ordinance allowing the removal of wolves that had a high probability of harming people. Following Catron County's lead, Otero County and Luna County enacted ordinances that prohibited the release of Mexican wolves into their counties. The FWS acknowledged that these actions violated the ESA.[3]

The counties' efforts received support in Congress. Representative Steve Pearce (R. N.M.) offered an amendment in the 2008 Department of Interior's appropriation bill to limit the Mexican wolf's range, to prohibit the reintroduction of additional wolves, and to curtail the funding for the program. The amendment failed, but Representative Pearce vowed to continue his efforts, stating: "The FWS must take active steps to better manage problem wolves and guarantee that farmers, ranchers, their families and their livestock are not repeatedly stalked and attacked."[4]

Environmental groups brought suit opposing the CCC ordinance. The U.S. District Court for the District of New Mexico dismissed the suit after the CCC amended the ordinance to remove the provision that allowed the unilateral trapping or removal of wolves from the wild.[5]

While the suit was underway, the CCC requested the removal of a female wolf (F294) that was guilty of three livestock depredations. After F294 was killed, Governor Richardson called for the suspension of SOP 13 in July 2007. SOP 13 had resulted in the removal of forty wolves from the BRWRA over the past three years, which was two times more than in the previous seven years.[6]

The Wild Earth Guardians (WEG) and the Rewilding Institute (RI) in April 2008 brought suit in the U.S. District Court for the District of Arizona, alleging the FWS and FS violated section 10(j) of the ESA because SOP 13 did not further Mexican wolf conservation. Furthermore, the FWS had abandoned its legal duty by delegating Mexican wolf management to the AMOC. The Defenders of Wildlife (DOW), representing ten other environmental groups, filed a similar suit. Both cases were consolidated in July 2008 as DOW v. Tuggle.

Environmental groups voiced their concerns. The DOW stated: "If management practices continue as they have under the AMOC, we could see the second extinction of the Mexican wolf in the wild. . . . Clearly FWS needs to take ultimate responsibility for the program and make wolf conservation a priority." The Center for Biodiversity (CBD) noted: "Transferring recovery responsibilities to an unaccountable bureaucracy that answers to the livestock industry has resulted in an eradication program masquerading as a recovery program."[7] The FWS, however, had already abandoned SOP 13 in 2008.

4.2 The Barack H. Obama Administration

Barack H. Obama took office in 2009. President Obama won New Mexico and lost Arizona in both the 2008 and 2012 presidential elections. The Obama administration put in place policies that increased and decreased livestock industry opposition to Mexican wolf recovery. In 2009, the FWS began a supplementary/diversionary feeding program for Mexican wolves. The supplementary feeding provided extra nutrition to enhance the wolf's survival and the diversionary feeding to decrease depredation by keeping the wolves away from livestock. The administration also supported the Wolf-Livestock Loss Demonstration Project Grants, which provided funds through a competitive process to eligible states and tribes to assist livestock producers in implementing nonlethal activities to decrease the possibility of wolf depredation and compensate them for losses resulting from wolf depredation. The administration completed the border barrier construction that the Bush administration had under contract. Nevertheless, the litigation regarding Mexican wolf recovery continued.

4.2.1 Defenders of Wildlife v. Tuggle

The Obama administration negotiated a settlement agreement in Defenders of Wildlife v. Tuggle in December 2009.[8] The FWS agreed to restructure the program. SOP 13 was formally ended and the AMOC was disbanded. The FWS noted "the parties that are part of the AMOC don't have the wolf's best interest at heart as does the FWS." Under the AMOC's management the Mexican wolf population which began at fifty-five in 2003 had decreased to forty-two by 2009. The FWS resumed control over the recovery program. Ranchers were not happy with the agreement, claiming huge economic losses from wolf depredation. The New Mexico Farm Livestock Bureau stated: "The FWS is just bending over backward to kowtow to these environmental groups. The FWS has zero credibility among ranchers. None of them want this."[9]

The Americans for the Preservation of Western Environment (APWE), joined by several counties and ranchers,[10] brought suit against the NMGFD and FWS challenging the changes in the recovery program, specifically the FWS's takeover of the program and the termination of SOP 13. The groups alleged the FWS "arbitrarily determines which management methods to implement and which to ignore" and asserted that a new Environmental Impact Statement (EIS) was required. The group's attorney stated: "The bottom line is that the individual landowners and small rural communities that are located in places in close proximity to where the wolf release program is being operated are not getting an adequate voice into the process."[11] The DOW and CBD intervened in the suit.

In December 2010, the U.S. District Court for the District of New Mexico in Americans for Preservation of Western Environment v. U.S. Fish and Wildlife Service dismissed the complaint against the NMGFD finding the plaintiffs failed to allege sufficient facts to support their claim. The plaintiffs withdrew their complaint against the FWS in February 2011. The CBD commented: "This lawsuit was entirely without merit. . . . With only a few dozen Mexican wolves struggling to survive in the wild, a new rash of federal trapping and shooting, would push this unique animal even closer to extinction."[12] At the end of 2011, there were sixty-seven Mexican wolves in the BRWRA.

4.2.2 Wild Earth Guardians v. Lane

Problems with New Mexico continued. Six months before leaving office, Governor Richardson issued an executive order that temporarily banned trapping in the BRWRA. Governor Richardson noted: "The indiscriminate traps and snares in the Recovery Area are harming efforts to reintroduce the Mexican wolf to its native habitat. I am ordering this temporary ban to protect the wolves and increase the likelihood for the wolves to survive and flourish."[13]

There had been six confirmed and three probable Mexican wolves trapped in New Mexico's portion of the BRWRA since 2002. Five wolves had been injured by traps, two severely enough to require leg amputations. Governor Richardson asked the NMGFD to assess the impact of trapping on the Mexican wolf. Following the executive order, the New Mexico Game and Fish Commission (NMGFC) extended the trapping ban for six months. The NMGFC, all appointed by the governor, developed wildlife policy, which was implemented by the NMGFD. Trapping had already been banned on public lands in Arizona and Colorado. The NMGFD commissioned the U.S. Geological Survey (USGS) to conduct a study regarding the impact of trapping on the Mexican wolf in the BRWRA.

Republican governor Susana Martinez took office in 2011, but did not share her predecessor's concern for Mexican wolf recovery. Governor Martinez ended New Mexico's participation in the recovery program in 2011. Representative Pearce applauded the decision.[14] The NMGFC declared it would no longer be involved in Mexican wolf management, but expected the FWS to continue the federal program. New Mexico had invested $507,644 over the past thirteen years. Federal funds received by the state over this time amounted to $1.4 million, constituting a $1.9 million investment in Mexican wolf recovery.[15] The abrupt exit was contrary to state law, which treated the Mexican wolf as an endangered species, requiring the state to support recovery.

New Mexico's departure from the 2010 MOU generated similar actions by Grant and Sierra Counties in New Mexico and the New Mexico Department of Agriculture. This left the AZGFD, the White Mountain Apache Tribe, and the Greenlee, Navajo, and Graham Counties in Arizona as the only remaining partners along with federal land and wildlife agencies.[16]

Environmental groups criticized New Mexico's action. The CBD noted: "We are worried that without New Mexico's help, wolves will prey on untended cattle in Gila National Forest and the livestock industry that demanded withdrawal of the state from management will then demand the trapping and shooting of the hapless wolves." Hunters supported the move. The Southwest New Mexico Outfitters countered: "Hunting is under fire from every front. It's not about the endangered wolf, or the spotted owl." Environmental groups use the ESA "very successfully as a tool to remove humans from the land. They view the wilderness as something aesthetic, like a utopian playground. They want humans out of those areas." The CCC declared: "So many ranchers are fed up with losing cattle. And we don't want these federal people on our land."[17]

New Mexico officials began to reconsider the trapping ban. The small business task force, all appointed by Governor Martinez, concluded that local economies would be "enhanced" by removing the trapping ban in wolf territory. The WEG petitioned the FS and the FWS for emergency exclusions in the Mexican wolf recovery area. The FS rejected and the FWS ignored the request.[18]

The USGS study commissioned by the NMGFD regarding the impact of trapping in the BRWRA was released in August 2011. The study noted that there had been seventy-eight Mexican wolf deaths since reintroduction, 80 percent caused by humans. Trapping, however, was only responsible for 4 percent of the deaths. The study found thirty-seven wolves were shot illegally, twelve were hit by vehicles, eleven were lethally removed, one was legally shot by the public, and one died of a trapping injury incurred in government research. There were thirteen trapping incidents in New Mexico: seven wolves were injured, two suffered leg amputations, and two died. The WEG stated: "The fact that two wolves died in traps was enough to make our jaws drop." The NMGFD Director Lane noted: "What it really boils down to is an emotional and social argument on whether trapping is indeed a management tool that should be allowed or not. Trapping is a valuable management tool for me."[19] Following the report, the NMGFC voted to end the trapping ban.

Environmental groups brought suit, arguing that with only forty-two Mexican wolves in the wild, every wolf death was significant. Numerous hunting, agriculture, and livestock groups intervened in the suit.[20] The AZGFD filed an amicus brief, supporting New Mexico's decision. The U.S. District Court for the District of New Mexico in Wild Earth Guardians v. Lane dismissed the suit and upheld lifting the trapping ban in December 2012.

The district court's analysis on various points was dubious. The federal district court held that section 9 of ESA, which prohibits private taking of endangered and threatened species, did not apply to nonessential experimental populations. Section 9 only applied to private taking of endangered and threatened species.[21] The district court, however, failed to recognize that the Mexican wolf, which was reintroduced as an experimental population, was considered to be a threatened species under the ESA and the FWS regulations, so it was protected under section 9(g) of the ESA, which prevents any person "to attempt to commit, solicit another to commit, or cause to be committed any offense defined in this section."[22]

The district court held that the NMGFD did not violate section 10(j) by allowing trapping in the BRWRA.[23] The district court, however, did not recognize that the section 10(j) rule only permitted an "unavoidable and unintentional take . . . which occurs despite reasonable care, is incidental to an otherwise lawful activity, and not done on purpose." The rule stipulated that "taking a wolf with a trap, snare, or other type of capture device within occupied wolf range . . . will not be considered unavoidable, accidental, or unintentional take, unless due care was exercised to avoid taking a wolf."[24] Between March 2002 and February 2009, fourteen wolves were captured fifteen times in foothold traps set by unauthorized personnel. Most were trapped within the BRWRA. Since the beginning of the program in 1998, the NMGFD has sold 21,734 resident trapping licenses and 313 nonresident trapping licenses that allowed individuals to place various types of traps in New Mexico to capture furbearers. These traps met state regulations, but could also take Mexican wolves. New Mexico never defined nor required trappers to exercise due care to avoid taking Mexican wolves.[25]

The district court also determined that the NMGFD could not be vicariously liable for harm to wolves by independent third parties.[26] The district court did not acknowledge that the state was responsible for issuing trapping licenses. No trap could be set in New Mexico without a license. The issuance of the license by the state was the "but for" causation resulting in the unlawful taking of Mexican wolves. Federal courts have found states liable for licensing activities by third parties that harm endangered and threatened species.[27] There were eighty Mexican wolves in the BRWRA in 2012.

4.2.3 New Mexico Game and Fish Department v. Department of Interior

New Mexico and Arizona continued to constrain Mexican wolf recovery. New Mexico left the recovery program in 2011, but allowed the release of Mexican wolves into the state until 2015. At the same time, the AZGFD decided to continue both its financial and infrastructure support for recovery but would not allow any new releases until a new recovery plan and new section 10(j) management rules were in place. The AZGFD amended its policy in 2012 to allow the release of replacement wolves for wolves lost due to an unlawful act or any other mortality factor. Environmental groups implored the FWS to release more Mexican wolves into the wild.[28]

In 2014, the NMGFC, all members appointed by Governor Martinez, instituted a new rule that required the NMGFD to approve all permits before allowing the release of wolves into the state, subject to concurrence by the NMGFC. In May 2015, the NMGFD rejected Ted Turner's Ladder Ranch application for a permit to import Mexican wolves for captive breeding. For the past seventeen years, the ranch had acted as a holding facility, only one of three, for wolves that would be eventually released into wild. The NMGFD noted that Turner could reapply later after a revised recovery plan was in place.[29] In August 2015, the AZGFD continued to prohibit the release of adult wolves but allowed cross-fostering, which involved placing captive-born pups into a wild litter, hoping the den mother would accept the pups as her own.

In January 2015, the FWS changed its regulation to allow Mexican wolves to be released directly from captivity into New Mexico. Previously captive wolves could only be released into Arizona, then disperse into New Mexico. In April and May 2015, the FWS requested two separate permits to release twelve Mexican wolves into Gila National Forest in New Mexico, two adults and ten pups for cross-fostering. The NMGFD denied both applications because the FWS had not submitted a revised wolf recovery plan with the request. The FWS appealed to the NMGFC, which upheld the NMGFD decision.[30]

The FWS informed the NMGFD that it would not comply with the state's permitting requirements and planned to reintroduce wolves without the state's permission. New Mexico was interfering with the FWS's legal obligation to recover the species by stalling the introduction of more genetically diverse captive wolves into the wild population. Environmental groups, supporting the FWS decision, declared that more releases into the wild were essential to diversify the gene pool and avoid inbreeding. The Obama administration had only released four Mexican wolves; three died and one was returned to captivity.[31]

In the spring of 2016, the FWS cross-fostered two nine-day-old Mexican wolf pups into a wolf den in New Mexico. Reactions were mixed. Environmental groups remained skeptical. The DOW noted: "The state is playing politics with this important animal. Unfortunately, the cross-fostering is not going to be enough to save the wolves in the wild. They need to introduce adult wolves as well." The NMGFD declared: "The FWS decision to move forward with unpermitted and illegal releases of wildlife into New Mexico demonstrates a disregard for our state's sovereignty. We have a responsibility to uphold the laws of our state and ensure the FWS complies with applicable state and federal law." The New Mexico Cattlemen's Association stated: "The federal government is turning out a predator to steal private property without compensation."[32]

In May 2016, the NMGFD brought suit, seeking a preliminary injunction to remove the two cross-fostered pups and halt further reintroductions without state approval. In order to issue an injunction, the court must assess (1) the likelihood of the plaintiff's success on the merits; (2) whether the plaintiffs will likely suffer irreparable harm; (3) whether the balance of equities tips in the plaintiff's favor; and (4) whether the injunction is in the public interest.[33] Environmental groups intervened to support the FWS decision. The DOW stated: "New Mexico's politically motivated lawsuit is a meritless, obstructionist attempt to usurp the FWS authority in endangered species recovery, as provided for in the ESA." The CBD noted: "The two captive born pups . . . embody the hope to diversify the Mexican wolf gene pool and save their kind from extinction. Removing these pups would be cruel and would contribute to an ongoing decline in wolf numbers and genetic diversity."[34]

The U.S. District Court for the District of New Mexico found in favor of the NMGFD on all four factors and granted the injunction halting further reintroductions. The court did not, however, order the FWS to remove the two reintroduced pups.[35] The district court's conclusions regarding three of the four factors were mistaken. The court's error concerning irreparable harm is discussed later.

4.2.3.1 Likelihood of Success on the Merits

The district court held that the NMGFD was likely to succeed on the merits of the case. Section 6(a) of the ESA requires the secretary to "cooperate to the maximum extent possible with the States." The district court determined that the "permissive language" in section 10(j), stating the secretary "may" authorize the release of an experimental population, did not create "a specific statutory directive requiring the Secretary to take action." Section 10(j) was simply a "grant of authority," not a specific mandate of the ESA. The importation and release of Mexican wolves was a discretionary act. The NMGFD refusal to issue a permit allowing importation or release of Mexican wolves did not interfere with interior's statutory duty.[36]

The district court's holding of the likelihood of success on the merits violated the "Chevron Doctrine," which held the federal courts "may not substitute their own construction of a statutory provision for a reasonable interpretation" by the agency. The Supreme Court in Chevron v. NRDC developed a two-step process regarding judicial review of an agency's legal interpretation. First, the court must determine "whether Congress has directly spoken to the precise question at issue."[37] This required the court to examine the text, intent, and purposes of the statute.[38]

The district court's decision was inconsistent with the text of section 6(a) of ESA, which only requires the secretary to "cooperate to the maximum extent practicable with the States."[39] It did not grant the states a veto over federal action.

The district court's decision was not supported by the legislative history. The reintroduction of experimental populations pursuant to section 10(j) did not require prior state approval. Congress recognized the importance of state involvement in reintroduction efforts, but did not indicate that the states would be the final arbiters of whether a release would occur. The 1982 House Report stated that section 10(j) "regulations should be viewed as an agreement among the Federal agencies, the state fish and wildlife agencies and any landowners involved. Changes in the regulations should only be made after close consultation with all of the affected parties."[40] The 1982 House Conference Committee stated that section 10(j) "obliges the Secretary to issue such regulations as he deems necessary and advisable to provide for the conservation of the experimental population." The secretary was granted "broad flexibility in promulgating regulations to protect such species."[41]

The district court's decision was inconsistent with the express purpose of the ESA "to provide a means whereby the ecosystems upon which endangered species and threatened species depend may be conserved [and] to provide a program for the conservation of such endangered species and threatened species."[42] Conservation means "to use and the use of all methods and procedures which are necessary to bring any endangered species . . . to the point at which the measure provided pursuant to [the ESA] are no longer necessary."[43]

The district court confused conservation (recovery) and survival (extinction). Section 10(j) authorizes the secretary to release the nonessential experimental population, which would contribute to the species' conservation (recovery) in the wild. The designation simply meant there was no danger of extinction if the nonessential population was lost in the wild. The ESA is concerned with both conservation (recovery) and survival (extinction). The reintroduction of the Mexican wolf was designed to achieve conservation (recovery) of the wolf in the wild, so that ESA protections could be removed.[44]

If Congress did not address the issue, the court could "not simply impose its own construction on the statute." The court must move to Chevron step two and determine "whether the agency's answer is based on a permissible construction of the statute." The court did not have to conclude that the agency's construction was "the only one it permissibly could have adopted to uphold the construction, or even the reading the court would have reached if the question had arisen in a judicial proceeding." Instead the court must defer to "a reasonable interpretation made by the administrator of the agency."[45]

Courts generally applied the "hard look doctrine" in the second step to determine if the agency decision was reasonable. The "hard look doctrine" requires the court to examine the agency's action "to satisfy itself that the agency has exercised a reasoned discretion, with reasons that do not deviate from or ignore the ascertainable legislative intent."[46] The FWS's decision to proceed with recovery in the absence of state permits was reasonable. The FWS was in the process of developing a revised recovery program, which entailed certain procedural steps. Furthermore, requiring the FWS to wait until the new recovery plan was formulated to obtain the necessary permits prevented the FWS from carrying out its statutory duty. More Mexican wolves were needed to realize recovery, particularly to improve genetic diversity. Otherwise, inbreeding would increase and the wild population would face extinction.

4.2.3.2 Balance of Equities

The district court held that "the balance of equities weighs in favor of issuance of the preliminary injunction." The short-term delay would not harm Mexican wolf recovery. Interior must comply with its regulations and state law.[47] The district court was wrong. The balance of equities favored the FWS.

The NMGFD argued that the current Mexican wolf population, 97 Mexican wolves, was close to the initial recovery goal, so additional wolves were not immediately needed.[48] The NMGFD failed to acknowledge that the recovery goals changed from 100 to 300–25 wolves in the revised section 10(j) regulations in 2015.

The FWS acknowledged that the "prime objective" of the 1982 recovery plan was to ensure the immediate survival of 100 Mexican wolves. The FWS explained that the population goals set in the 1998 regulation, which were consistent with the 1982 recovery plan, were no longer adequate to protect the Mexican wolf. The FWS acknowledged that it could not "achieve the necessary population growth, distribution, and recruitment that would contribute to the persistence of, and improve the genetic variation within, the experimental population under the current regulations." Mexican wolves might "disperse into southern Arizona and New Mexico from the reintroduction areas in the States of Sonora and Chihuahua in northern Mexico." Certain management provisions were inadequate. Furthermore, the change in status of the Mexican wolf from the generic gray wolf population listing to an endangered subspecies in 2015 necessitated the rule change.[49] Additional wolves must be introduced to conserve the subspecies.

The NMGFD argued that there was no danger of inbreeding and genetic deterioration in the small population of Mexican wolves.[50] The FWS, however, noted that the number of initial releases with a better success rate was necessary to improve the genetic fitness of the experimental population. Higher levels of genetic variation were crucial to diminishing the risk of inbreeding and ensured the persistence of the population, particularly until the population reached at least 250.

The FWS pointed out that the genetic diversity of the Mexican wolf population was problematic. In nineteen of the twenty-one potential breeding pairs for 2016, one of the breeding pairs would be a descendant of the Bluestem pack. In eleven of the twenty-one (52 percent) potential breeding pairs, both individuals would be descendants from the Bluestem pack. Field observations suggested an even greater kinship. Furthermore, thirty-seven of the forty-two (88 percent of the individuals with known genetics) were related to the Bluestem pack, indicating that related pair matching in the future was high. This could have long-term implications because the breeding of close relatives aggravates genetic problems.[51]

4.2.3.3 Public Interest

The district court determined that the state's interest in managing its wildlife was the paramount public interest. Mexican wolves must be managed or they would cause environmental havoc. Wolves could become a public nuisance. The FWS must comply with state law and obtain a permit before importing or releasing Mexican wolves into New Mexico.[52]

The district court decision was contrary to the ESA policy to protect ecosystems. The 1982 House Conference Committee declared that "individual species should not be viewed in isolation, but must be viewed in terms of their relationship to the ecosystem of which they form a constituent element." Even though the ESA focuses on endangered and threatened species "the purposes and policies of the Act are far broader than simply proving for the conservation of individual species or individual members of listed species."[53]

The Mexican wolf plays an important role in maintaining ecosystem balance. Plants and animals exist in interconnected ecosystems. The loss of one species affects the entire system. Disruptions create environmental instabilities that diminish nature's ability to establish food chains, cycle nutrients, maintain the quality of the atmosphere, control the climate, regulate the fresh water supply, maintain the soil, dispose of wastes, pollinate crops, and control pests and disease.[54] Senator Tunney (D. Cal.), the floor leader and member of the conference committee regarding the ESA, declared that each species provides a service to the environment and is part of a complex ecosystem which depends on all its components for stability. The value of each species is unknown, so its loss can't be assessed.[55]

Robert Costanza estimated the value of ecosystem services to be in the range of $16 to $55 trillion. With an estimated annual value of $33 trillion per year, ecosystems provide services that cost almost twice the total gross national product of all the nations of the world combined. Costanza noted that ecosystem services are not fully "captured" in commercial markets, so they are given little weight in policy decisions. This may jeopardize human existence. Costanza warned that "the economies of the Earth would grind to a halt without the services of ecological life-support systems, so in one sense their total value to the economy is infinite."[56] Robert Costanza's updated 2011 study, utilizing the same study parameters, concluded that ecosystem services provide benefits worth between $125 and $145 trillion per year.[57]

The Mexican wolf plays important role in preserving biodiversity. Biodiversity is the "total of genes, species and ecosystems on the earth." Biodiversity is a "living exploitable renewable resources," which has "economic importance and potential consumptive and transformative uses." The preservation of genes is important for the development of food and medicine and the maintenance of the ecosystem. Companies seek genetic material.[58]

The ESA is concerned with the preservation of biodiversity. The 1973 House Committee Report on the ESA states, "the value of endangered species is quite literally incalculable. . . . From the most-narrow possible point of view, it is in the best interest of mankind to minimize the loss of genetic variations. The reason is simple: they're potential resources. They are the keys to puzzles, which we cannot solve, and may provide answers to questions which we have not yet learned to ask."[59]

Senator Tunney pointed out that each species is important for science. The diversity of genetic types is necessary for thorough scientific knowledge. The unknown potential of investigation into genetic structure must remain unhindered to produce knowledge for the benefit of man.[60]

Federal courts have recognized the importance of biodiversity.[61] The Supreme Court in TVA v. Hill acknowledged congressional concern "about the unknown uses that endangered species might have and about the unforeseeable place such creatures may have in the chain of life on this planet."[62]

The benefits of Mexican wolf restoration are spread across the ecosystem. A wide variety of scavengers and other carnivores acquire sustenance from carrion being readily available year-round rather than one-time in the early spring because of winter deaths. When wolves make a kill, sustenance is provided for the entire food chain. After wolves are finished eating, other carnivores and scavengers take their share, insects clean the carcass, and birds feed on the insects. In addition, wolf predation improves soil nutrients, soil microbes, and plant quality by increasing deposition and distribution of prey carcasses over the landscape.[63]

The Mexican wolf also maintains an important balance among predators. The Mexican wolf is a summit predator that keeps smaller predators (mesopredators), like the coyote, in check. A 2009 study pointed out that in the absence of summit predators, "mesopredator outbreaks often lead to declining prey populations, sometimes destabilizing communities and driving local extinctions." When wolves disappear, they are replaced by coyotes, which kill thousands of sheep all over the West. Coyote populations occur at higher densities and are more adaptable, so they pose a greater threat to livestock. Given the high cost of controlling mesopredators, returning summit predators, like the wolf, would be cheaper and more effective. Furthermore, diminution of the coyote leaves much of the coyote's prey, mainly small rodents, for predatory birds, such as hawks, eagles, and owls. Red foxes and bobcats also benefit from reduced coyote competition for food, particularly during the winter months.[64]

Similar findings in a 2017 University of Washington study concluded the range of top predators, like the Mexican wolf, is so diminished and fragmented that they cannot control small predators. Aaron Wirsing, a coauthor of the study, stated: "It will require managing for top predator persistence across large landscapes, rather than just in protected areas, in order to restore natural predator-predator interactions." Furthermore, he noted that "coyotes have essentially hitched a ride with people. Not only do we subsidize coyotes, but we also helped them by wiping out their predators: wolves."[65]

Mexican wolves play an important role in regulating prey population, which include elk, white-tailed and mule deer, and to a lesser extent pronghorn, javelina, and Rocky Mountain bighorn sheep. Elk is the preferred prey and constitutes the majority of their diet. In the absence of predation, prey populations increase to and often above the carrying capacity of their environment. This causes resource scarcity, resulting in poor nutrition that adversely affects the prey population's health. Reducing the prey population below carrying capacity allows for a positive annual increase in the population.

Mexican wolves also benefit their prey in other ways. Wolves are selective hunters and usually choose more vulnerable and less fit prey. Younger and older animals and diseased and injured animals are taken in greater proportion than healthy, prime-aged individuals. Wolf predation can suppress the emergence of disease in ungulates and limit its prevalence, in part by reducing density and group sizes of the elk and deer. This reduces or eliminates the spread of brucellosis and chronic wasting disease. In addition, causing ungulates to graze in smaller groups potentially slows the spread of ungulate diseases that persist among high-density populations of ungulates.[66]

The presence of the Mexican wolf changes prey behavior. Ungulate prey avoids areas where wolves are present, reducing adverse impact on plants. Undisturbed browsing by large ungulates is an important factor affecting biodiversity and plant communities. Increased browsing affects the growth and regeneration of trees and shrubs. Any decrease in prey or concentrated herbivory restores the vegetation. It also relieves pressure on riparian corridors, so there is less erosion. This affects water quality, prevents flooding, and provides opportunities for beavers.[67]

Environmental groups criticized the decision. The DOW noted: "the court's decision today takes the Mexican wolf one step closer to extinction in the wild. . . . The states are playing puppy politics and working to tie the hands of the FWS in their duty to recover the lobo." The CBD declared: "This shocking ruling will hopefully not stand on appeal. We respectfully believe that the court misread the ESA intent, but unfortunately, while the ruling stands the plight of the Mexican wolf continues to worsen."[68]

4.2.4 Tenth Circuit

Interior appealed the decision in the Tenth Circuit. Seventeen states filed an amicus brief supporting the NMGFD, asserting that reintroducing wolves would have a devastating impact on wildlife management and livestock production. The Tenth Circuit overturned the district court. The Tenth Circuit held the NMGFD did not demonstrate that it would suffer any irreparable harm, but did not address the other three factors.[69]

4.2.4.1 Irreparable Harm

The district court held the NMGFD would suffer irreparable harm because it would not be able to comprehensively manage wildlife within its borders. The NMGFD managed ungulate herds to determine how many hunting licenses to issue. Predator and prey must be managed together. The NMGFD's lack of knowledge regarding the numbers, location, and timing of the release of Mexican wolves, an apex predator, would disrupt its efforts to manage wildlife. This disruption could not be compensated by money.[70]

The Tenth Circuit reversed, stating that the district court "failed to establish a significant risk of irreparable injury in the absence of an injunction."[71] The Tenth Circuit recognized that the NMGFD knew the time, manner, and location of the release of Mexican wolves. The wolves could be tracked because prior to release they were marked and fitted with radio collars.[72]

The NMGFD's assertions of irreparable harm were based on speculative conclusions. The NMGFD asserted that a single Mexican wolf kills over twenty elk or deer per year. The presence of 300 to 325 Mexican wolves would have a major impact on the state's ungulate population. The NMGFD exaggerated the potential harm to its wildlife. The FWS only planned to release one or two packs, two adults and several pups, every four years for the next eight years, one or two packs during the following three successive generations, until year twenty in the program. Furthermore, the success rate for releases from 1998 through 2013 was only 21 percent. For every 100 Mexican wolves released only twenty-one survived. The FWS projected that the Mexican wolf population would not reach 300 wolves for at least thirteen years.[73]

The FWS concluded that the density of wolves in the recently established Mexican Wolf Experimental Population Area (MWEPA) with a population of 300 to 325 was expected to be no higher (more likely lower) than the current level because the wolf recovery area was larger. The wolf/elk ratios, which are an indicator of predation pressure, were also expected to occur at levels that would not cause significant biological damage and not adversely affect the ungulate population. The wolf/elk ratio in 2014 was 2.56 wolves for every 1,000 elk. This would only increase to 3.9 wolves for every 1,000 elk with the 300 to 325 wolf population. Furthermore, Mexican wolves tended to prey on unproductive calves and older cows, leaving the remaining elk for hunters. Hunter visitations and success rates since 1998 in areas occupied by the Mexican wolf were stable. The FWS forecast that these trends would continue during the project's time period.

The FWS cited the AZGFD analysis of data from 1998 through 2012 regarding the Mexican wolf's impact on the ungulate population in the BRWRA. The FWS noted that Mexican wolves target elk as their primary prey, particularly calves during the spring and summer season. Nevertheless, the number of elk calves that survived through early fall remained constant. There was a similar finding regarding mule deer. Furthermore, the final rule did allow for the taking of wolves if they caused unacceptable impacts on wild ungulate herds.[74]

4.2.4.2 State Sovereignty

On appeal, the NMGFD argued that any release of Mexican wolves would significantly interfere with the core government function to enforce its laws within its borders. This interference with state sovereignty constituted irreparable harm. The district court did not address this issue, but the Tenth Circuit rejected the NMGFD's claim, stating: "The Department has not presented any factual or legal basis for finding that FWS's anticipated releases would interfere with the State's ability to establish or enforce its laws, or that the releases would pressure the State to change its laws."[75]

The NMGFD claim implicitly rested in part on the old state ownership of wildlife theory. The Supreme Court in Geer v. Connecticut upheld a state statute that prohibited the export of game birds. The Court declared the state could "control and regulate the common property in game" because it holds such a right in "trust for the benefit of the people."[76] The state ownership theory was overturned in 1979 in Hughes v. Oklahoma. The Court invalidated an Oklahoma statute that prohibited the export of natural minnows and declared that the Geer analysis had eroded "to the point of virtual extinction in cases involving the regulation of wild animals." The Court noted that "the ownership language . . . must be understood as no more than a nineteenth century fiction expressing the importance to its people that a State have power to preserve and regulate the exploitation of an important resource." Wildlife regulation must be evaluated according to the same principles applied to state regulation of other natural resources. State power only extended as far as its exercise was not incompatible with, or restrained by, the constitutional rights conveyed to the federal government. Nevertheless, the court acknowledged that there were legitimate state concerns regarding the conservation and protection of wild animals within their borders.[77]

Reactions to the decision varied. The NMGFD stated: "While we're disappointed in the court's ruling, the case will continue to move forward. We'll continue to do all we can to show how unpermitted, experimental release of Mexican wolf by the federal government will be harmful to New Mexico." The CBD declared: "The ruling is noteworthy in pointing out that the Martinez administration was unable to articulate any harms from the wolves. The decision makes clear that the FWS has the authority to do what's needed to save the Mexican gray wolf and other endangered species from extinction." The WEG noted: "Today's ruling is monumental in upholding the integrity and ambition of the ESA. We are encouraged by the court's refusal to allow a minority of anti-carnivore political interests to hold the endangered lobo hostage and urge the Service to immediately resume much needed releases so that the Mexican wolf may truly recover."[78]

4.3 Post-Litigation Changes

Even though the FWS prevailed in the NMGFD litigation, Governor Martinez continued to play politics. After the decision, the NMGFD in May 2017 issued a permit allowing the cross-fostering of two captive-born pups into a wolf den. However, the NMGFD demanded that for each pup introduced, another wild-born pup had to be removed from the den and placed in captivity. The NMGFD did not want any increase in the wolf population, so the FWS was forced to make "Sophie's Choice."[79]

Environmental groups criticized the policy. The DOW stated: "This is an absurd condition and the antics coming out of the state of New Mexico are inexcusable. The state of New Mexico should let the scientists at the FWS do whatever is necessary to help rebuild the wild population of the world's most endangered gray wolf species." The CBD noted that the FWS "should not agree to remove wolves. This is a recovery effort and the numbers should be going up." Nevertheless, the FWS capitulated. Two wolf pups were cross-fostered, but the two others removed from the den were placed in captivity.[80]

After Democrat Michelle Lujan Grisham defeated Republican governor Martinez and took office in 2019, New Mexico rejoined the Mexican wolf recovery program, which it had left in 2011. New Mexico signed the 2020 MOU, which revised the 2015 MOU to address the 2015 changes in management rules and 2017 recovery plan.[81] In addition, there was an effort to end trapping on public land in New Mexico.[82]

Conclusion

Public law litigation during the Bush and Obama administrations had mixed results. The settlement agreement in Defenders of Wildlife v. Tuggle promoted changes in the program. The FWS terminated the AMOC and ended SOP 13, which had resulted in the killing of too many valuable Mexican wolves. The FWS resumed primary management authority over the recovery program.

Public law litigation was not always successful. The federal district court in Wild Earth Guardians v. Lane upheld New Mexico's policy allowing trapping in the BRWRA. Judicial skeptics would support the decision, which upheld state government policy. Judicial proponents would be critical because New Mexico's policy violated the ESA. The Mexican wolf population was low, so the loss of any wolf could jeopardize the success of the program.

In the next battle, the federal district court in New Mexico Game and Fish Department v. U.S. Fish and Wildlife Service upheld New Mexico's refusal to permit the release of Mexican wolves into the state until a new recovery plan was adopted. The Tenth Circuit reversed the district court decision, which undermined federal statutory and constitutional authority, and affirmed the FWS's primary authority over Mexican wolf recovery. This was important because the release of captive wolves was necessary to expand and increase the genetic diversity of the wolf population. Both judicial skeptics and proponents would concur with the Tenth Circuit decision, which upheld federal policy and carried out the ESA mandate.

While the NMGFD litigation was underway, there were substantial changes in the recovery program. The Obama administration changed the status of the Mexican wolf to an endangered subspecies. The FWS submitted revised section 10(j) management regulations and an updated revised recovery plan. These issues are addressed in the subsequent chapters.

Notes

1 "State Supports Wolf Reintroduction Program," Associated Press State & Local Wire (April 8, 2004).

2 Defenders of Wildlife, Complaint for Declaratory and Injunctive Relief, *DOW v. Tuggle*.

3 Sue Major Holmes, "Endangered Wolf Program Remains a Complex, Volatile Issue," Associated Press State & Local Wire (June 23, 2007). "Conservation Group Oppose Catron County Action against Wolves," Associated Press State & Local Wire (May 3, 2007).

4 "Pearce Disappointed with Vote on Wolf Recovery Program," State News Service (June 26, 2007).

5 Susan Montoya Bryan, "NM Judge Rules on Mexican Wolf Ordinance," Associated Press State & Local Wire (October 2, 2008).

6 "Wildlife Groups Call for End to Mexican Wolf Removal Policy," Associated Press (May 1, 2008).

7 Chris Kahn, "Wildlife Groups Call for End to Mexican Wolf Removal Policy," Associated Press (May 1, 2008). "US Wildlife Officials Failing to Conserve Mexican Wolf," Targeted News Service (May 30, 2008).

8 607 F. Supp. 2d 1095 (D. Az., 2009). Susan Montoya Bryan, "Judge Sides with Environmentalists in Wolf Case," Associated Press State & Local Wire (April 3, 2009).

9 "Wolves: FWS Agrees to Revamp Southwest Management Program," Land Letter (November 19, 2009). Susan Montoya Bryan, "Wolf Recovery at Cross Roads in Southwest," AP Online (December 6, 2009).

10 Gila National Forest Permittees Assoc., Catron and Otero Counties and Boards of Commissioners.

11 "NM Counties, Ranchers Sue to Halt Reintroductions," Land Letter (August 26, 2010).

12 "Livestock Growers, Counties Withdraw Lawsuit to Eradicate Gray Wolves From New Mexico," Targeted News Service (February 1, 2011).

13 "Governor Bill Richardson Orders Temporary Trapping Ban," Targeted News (July 28, 2010).

14 "Opponents of Wolf Recovery Reacted Favorably. Commission Suspends MW Reintroduction Program," Las Cruces Sun-News (New Mexico) (June 10, 2011). "Congressman Pearce Commends Game Commission on Wolf Decision," Congressional Documents and Publications (June 10, 2011).

15 "Mexican Gray Wolves Face New Challenges," Deseret Morning News (Salt Lake City) (July 11, 2011).

16 Susan Montoya Bryan, "Support Waning for Mexican Gray Wolf Program," Associated Press State & Local Wire (April 8, 2012).

17 "Commission Suspends Mexican Wolf Reintroduction Program," Las Cruces Sun-News (June 10, 2011).

18 "NM Game Officials to Consider Ending Trapping Ban," Associated Press State & Local Wire (July 21, 2011). "Pressure Resumes for Trapping Ban in Wolf Area," Associated Press State & Local Wire (July 30, 2011).

19 "Feds Release Study of Trapping Effects on Wolves," Associated Press State & Local Wire (August 8, 2011).

20 New Mexico Trappers Assoc., New Mexico Council for Outfitters and Guides, New Mexico Farm and Livestock Bureau, Coalition of Arizona/New Mexico Counties for Stable Economic Growth, United Sportsmen for Fish and Wildlife, New Mexico Cattlegrower's Association, New Mexico Woolgrowers, New Mexico Federal Lands Council

21 *WEG v. Lane*, CIV 12-118 LFG/KBM, December 4, 2012, 7–10.

22 16 U.S.C. 1538 (g) and 1539 (j)(C). 50 C.F.R. 17.83 (2016).

23 *WEG v. Lane*. 8–10,

24 50 C.F.R. 17.84 (k)(15).

25 *WEG v. Lane*, 7–8.

26 *WEG v. Lane*, 8–10.
27 *Strahan v. Coxe*, 127 F.3d 155, 164 (1st Cir. 1997).
28 "Plan Announced for Arizona Mexican Wolf Release in 2014," Plus Media (February 17, 2014).
29 N.M. Code R. sect. 19.35.7.8 and section 19.33.10.11. "Feds Push Game Commission to Allow Mexican Wolf Release," Santa Fe New Mexico (August 30, 2015).
30 "Feds Move Ahead with Mexican Wolf Releases," New Mexican (October 14, 2015).
31 43 C.F.R. section 24.4 (i)(5). "Wildlife Advocates, Scientists Call on Secretary Jewell to Hasten Release of Mexican Gray Wolf into New Mexico," Center for Biodiversity (October 8, 2015).
32 "Gray Wolf Pups Released into New Mexico Wild, Santa Fe New Mexican" (April 30, 2016). "Department Takes Actions to Stop FWS from Ignoring Federal and State Law," Targeted News (May 13, 2016).
33 *Prairie Band of Potawatomi Indians v. Pierce*, 253 F.3d 1234, 1246 (10th Cir. 2001).
34 CBD, DOW, WEG, New Mexico Wilderness Alliance, "Conservationists Intervene on Behalf of Mexican Wolf Reintroduction Efforts," Targeted News (June 6, 2016).
35 NMGFD v. Department of Interior, Memorandum Opinion and Order Granting Petitioners Motion for Preliminary Injunction and Proposed Order of Injunction (Memo), No. CV 16–00462 WJ/KBM (June 10, 2016), 1–2.
36 Memo, 17–18.
37 467 U.S. 837. 824–43 (1984).
38 *INS v. Cardozo-Fonseca*, 480 U.S. 446, 446–49 (1987).
39 16 U.S.C. 1535 (a).
40 H.R. Report 97-567, 34.
41 H. Conf. Comm. Rpt., 97–835, 29–30 (1982).
42 16 U.S.C. 1531 (b).
43 16 U.S.C. 1532 (3).
44 *Gifford Pinchot Task Force v. FWS*, 378 F.3d 1069, 1069–70 (9th Cir. 2004). *Sierra Club v FWS*, 245 F.3d 434, 441–2 (5th Cir. 2001).
45 467 U.S., 842–5.
46 Jonathan T. Molot, "Judicial Perspectives in the Administrative State," 53 Stanford Law Review 1, 92–9 (2000). *Greater Boston Television Corp. v. FCC*, 444 F2d 841, 850–1 (D.C. Cir. 1970).
47 Memo, 21.
48 Response Brief for Plaintiff-Appellee NMGFD, 34 (Response Brief).
49 80 Federal Register 2512, 2517–18, 2529 (January 16, 2015).
50 Response Brief, 34–5.
51 80 Federal Register 2488, 2504–6 (January 16, 2015).

52 Memo, 21–2.
53 House Conference Committee Report 97-835, 30 (1982).
54 Patrick Parenteau, "Rearranging the Deck Chairs: ESA Reforms in an Era of Mass Extinctions," 22 William & Mary Environmental Law & Policy Review 227, 236–44 (1998). Paul and Anne Ehrlich, *Extinction: The Causes and Consequences of the Disappearance of Species* (Random House, 1981), 86–95.
55 119 Congressional Record 25668 (July 24, 1973).
56 Robert Costanza et al., "The Value of the World's Ecosystem Services and Natural Capital," 387 Nature 253, 259 (1987).
57 Robert Costanza et al., "Changes in the Global Value of Ecosystem Services," 26 Global Environmental Change 152–8 (2014).
58 Mark A. Urbanski, "Chemical Prospecting, Biodiversity Conservation, and the Importance of International Protection of Intellectual Property Rights in Biological Materials," 2 Buffalo Journal of International Law 133, 134–5 (1995) Eric Christensen, "Genetic Ark," 40 Stanford Law Review 279, 285–8 (1987).
59 H.R. Rep. No. 412, 93d Cong., 1st Sess. (1973) 4–5.
60 119 Congressional Record 25668–70 (July 24, 1973).
61 *National Association of Home Builders v. Babbitt*, 130 F.3d 1041 (D.C. Cir. 1997). *U.S. v. Bramble*, 103 F.3d 1475 (9th Cir. 1996).
62 *TVA v. Hill*, 437 U.S. 153, 178–9 (1978).
63 FEIS, Proposed Revision to the Regulations for the Nonessential Experimental Population of Mexican Wolf (Canis Baileyi) ch. 4, 11–12 (2014) (FEIS).
64 Laura R. Prugh, et al., "The Rise of the Mesopredator," 59 Bioscience 779 (October 2009). "Goodbye Wolf, Hello Coyote and More," States News Service (January 15, 2010). FEIS at 17–18.
65 "Wolves Need Space to Roam to Control Coyote Populations," Targeted News (May 23, 2017), citing Thomas M. Newsome et al., "Top Predators Constrain Mesopredator Distribution," 8 Nature Communications 15469 (2017).
66 80 Federal Register, 2492, 2508, 2555.
67 William J. Ripple and Robert L. Bescheta, "Trophic Cascades in Yellowstone: The First 15 Years after Wolf Reintroduction," 145 Biological Conservation 205–13 (January 2012).
68 "Court Enjoins Release of Mexican Wolf in New Mexico," Plus Media (June 15, 2016). "Court Grants New Mexico Preliminary Injunction on Mexican Wolf Releases," State News Service (August 6, 2016).
69 *NMGFD v. Interior*, FWS v. DOW, 854 F.3d 1236 (10th Cir. 2017). (NMGFD)
70 Memo, 19–21.
71 NMGFD, 1247–54.

72 Amicus Curiae Brief of Foundation to Protect New Mexico Wildlife in Support of Department of Interior at 17–19.

73 NMGFD, 1253 citing 80 Federal Register 2512-01, 2524.

74 NMGFD, 1253. 80 Federal Register, 2555.

75 NMGFD, 1254.

76 161 U.S. 519, 534 (1896).

77 441 U.S. 322, 331–6 (1979).

78 "WEG: Major Victory for Lobo Recovery in Land of Enchantment," Targeted News (April 25, 2017). "Court Lifts Injunction on Mexican Wolf Releases in New Mexico," Albuquerque Journal (April 27, 2017).

79 Susan Montoya Bryan, "New Mexico Issues Permit to Allow for Fostering of Wolf Pups," Associated Press State & Local Wire (May 4, 2017). "State of New Mexico Issues Sophie's Choice," State News Service (May 5, 2017).

80 "New Mexico Issues Permit to Allow for Fostering of Pups," Associated Press State & Local (May 4, 2017). "Feds Release of Endangered Wolf Pups in New Mexico," Associated Press State & Local (May 5, 2017).

81 "NM Rejoins Pact to Restore MW," Idaho Statesman (October 25, 2019).

82 "Inconsistency at Game and Fish: After State Rejoins Wolf Recovery Program," New Mexico Political Report (December 10, 2019).

5

Mexican Wolf Management

One of the major issues confronting Mexican gray wolf recovery concerned the use of "the best scientific and commercial data available" (best available science) in the decision-making process as required by the Endangered Species Act (ESA). After the termination of the Adaptive Management Oversight Committee, the U.S. Fish and Wildlife Service (FWS) management of Mexican wolf recovery under section 10(j) continued to be criticized. This chapter analyzes the litigation in Center for Biodiversity v. Jewell dealing with the change in the Mexican wolf's status in the Obama administration; the litigation in Center for Biodiversity v. Jewell invalidating the revised section 10(j) management regulations in the Obama administration; and the litigation in Center for Biodiversity v. Haaland upholding the updated section 10(j) management regulations promulgated in the Biden administration.

5.1 Change in Status

Environmental groups petitioned the FWS in 2009 to change the status of the Mexican wolf to an endangered subspecies and establish its critical habitat. Environmental groups hoped this change would encourage the FWS to produce a new recovery plan that would allow the Mexican wolf to migrate into the Southern Rocky Mountains. In August 2010, the FWS announced that the petition presented substantial information indicating a Mexican wolf subspecies designation might be warranted.

At the end of a twelve-month study in October 2012, the FWS denied the petition because the species was already listed under the ESA.[1] At the time, the FWS was in the process of reviewing the status of the gray wolf nationwide. In 2013, the FWS, relying on an in-house study by Steven M. Chambers,[2] determined that there were three distinct species of gray wolves currently inhabiting the lower forty-eight states: *Canis lupus*, *Canis rufus*, and *Canis lycaon*. Canis lupus could be subdivided into three subspecies: *Canis nubilus*, *Canis occidentalis*, and *Canis baileyi* (Mexican wolf). The FWS planned to delist *C. I. nubilus* and *C. l. occidentalis*, but to grant *C. l. baileyi* (Mexican wolf) status as an endangered subspecies within the southwest Distinct Population Segment (DPS) across central and southern Arizona and New Mexico. After criticism from prominent scientists, the Obama administration withdrew the proposal.[3]

Environmental groups brought suit. The resulting settlement agreement in Center for Biodiversity v. Jewell required the FWS to reconsider its listing decision. The final rule designating the Mexican wolf as an endangered subspecies was published in January 2015. The FWS justified the change by citing illegal killings, inbreeding (mating of close relatives), the loss of heterozygosity (a decrease in the proportion of individuals in a population that have two different alleles for a specific gene), the loss of adaptive potential, the small population size, and the cumulative effects of these aforementioned threats. Furthermore, absent ESA protection, "regulatory protection would not be adequate to ensure the survival of the Mexican wolf."[4] The change in status required the promulgation of revised section 10(j) regulations. There were ninety-seven wolves in the BRWRA at the end of 2015. This was down from the 110 wolves in 2014, primarily because of low pup survival.

5.2 Revised Section 10(j) Management Regulations

In November 2012, the Center for Biodiversity (CBD) filed a complaint seeking to compel the FWS to conduct formal rulemaking to amend the section 10(j) regulation regarding Mexican wolf management. In August 2013, the U.S. District Court for the District of Columbia ordered the FWS to submit revised 10(j) regulations by January 2015.[5]

While the litigation was underway, efforts by House Republicans in Congress to frustrate Mexican wolf recovery continued, but were unsuccessful. Representative Denny Rehberg (R. Mt.) introduced a bill that stripped ESA protections from the gray wolf, including the Mexican wolf. Representative Steve Pearce (R. N.M.) attempted to attach a rider to the stopgap 2011 budget bill that ended the funding for Mexican wolf recovery. The amendment failed, but he cosponsored another bill that eliminated protection for all gray wolves under the ESA, silencing the lobo forever. Representative Paul Miller (R. Mich.) introduced a bill, which eliminated protections once the Mexican wolf population reached one hundred.

In January 2015, the FWS produced revised section 10(j) regulations, which provided for the following: First, there was a fourfold increase in the area that Mexican wolves could occupy and a tenfold increase in the area where Mexican wolves could initially be released from captivity. The BRWRA was terminated and replaced by the Mexican Wolf Experimental Population Area (MWEPA), which included all areas in Arizona and New Mexico south of I-40 to the Mexican border. The MWEPA was divided geographically into three zones, each designated for release, translocation, or dispersal of wolves. Mexican wolves could initially be released or translocated in Zone One, which included the Apache, Gila, and Sitgreaves National Forests, the Tonto Basin Ranger District of Tonto National Forest, the Magdalena Ranger District of Cibola National Forest, and the Arizona towns of Payson and Pleasant Valley. Wolves would be allowed to naturally disperse and occupy this zone. Mexican wolves would be allowed to disperse into, occupy, and be translocated to Zone Two. Pups under five months of age could also be released on federal land in Zone Two. Zone Three contained less suitable habitat where Mexican wolves would be more actively managed to reduce conflict with the public. The expanded MWEPA, unlike the BRWRA, which principally contained national forest land, included a significant amount of nonfederal land. Neither initial releases nor translocations would occur in Zone Three, but Mexican wolves would be allowed to disperse into and occupy this zone. There would be a phased approach to translocations, initial releases, and occupancy west of Highway 87 in Arizona. Wolves would not be allowed to migrate north of I-40 without Utah and Colorado permission.

Second, the population objective provided for 300 to 325 wolves after thirteen years and called for "the persistence of [the] population and . . . its removal from the endangered species list." The FWS decided "additional releases of more than one to two effective migrants per generation would address the inbreeding problems." A genetically effective migrant is defined as wolf that breeds outside its pack and produces at least one pup that survives until December 31 of the year of its birth. Furthermore, the FWS would exercise a wide range of management options, particularly translocation.

Third, the new rule modified circumstances for lethal and nonlethal takings to provide greater management flexibility and avoid conflicts with livestock owners and hunters. The rule focused on unacceptable impacts on wild ungulate herds. If the Arizona Game and Fish Department (AZGFD) or the New Mexico

Department of Game and Fish (NMGFD) determined wolves were having an adverse impact on wild ungulate herds, either department could request the removal of the wolves from the impacted area. The agencies would be required to "submit a science-based report, which was subject to peer review and public comment," explaining why the wild ungulate herd was below management objectives and the agency's efforts to halt the decline. An "unacceptable impact" would be determined by the state agency based on management objectives or a 15 percent decline in the ungulate herd documented by the state agency. If these requirements were met, the FWS would establish the management actions necessary for species conservation. Since reintroduction, there had been "no discernable impact" from Mexican wolf depredation of their prey in the BRWRA. The FWS further projected that wolves would have "little or no effect on the abundance of elk and deer across most of Arizona and New Mexico where elk and deer abundance is stable, or above population objectives." Finally, the revised rule maintained the nonessential experimental designation. The FWS determined that reconsideration of the population's nonessential status was beyond the scope of the rulemaking.[6]

The FWS asserted that the rule was just the "first step toward recovery." Expanding and improving the genetic health of the population would increase the possibility of species recovery. The FWS acknowledged that additional measures would be necessary in the future to recover the species, including objective and measurable criteria for delisting, a science-based population goal, and an expanded dispersal area based on establishing a metapopulation. The FWS planned to review progress under the new rule in the next five years.[7]

The revised rule was criticized by environmental groups on several grounds. First, Mexican wolves could still be killed on certain private and state land, even in the absence of livestock predation, if the wolves posed an "unacceptable impact" on big game. State managers had too much discretion to determine when wolves posed an "unacceptable impact" on prey species, triggering their removal. Second, Mexican wolves needed more room to roam. Much of the expanded area in the MWEPA contained unsuitable habitat. Third, the FWS ignored the best available science. The rule was contrary to the 2012 draft report, which asserted that wolves must be able to move north of I-40 into southern Colorado and Utah. The best available science called for three genetically linked subpopulations of 250 wolves each in the southwest. Fourth, the multiphase implementation approach restricted reintroduction for another twelve years.

Fifth, the FWS capitulated to state demands after the AZGFD threatened litigation.[8]

The NMGFD was also skeptical because the rule did not establish the number of wolves that would constitute recovery. The regulation did not focus on suitable habitat and failed to articulate what constituted the unacceptable impacts on wildlife. Furthermore, wolf recovery threatened energy development in the Permian Basin. The Defense of Free Enterprise, a wise use group, characterized the plan as a federal land grab, constituting a taking of property.[9]

5.3 Center for Biological Diversity v. Jewell

Environmental groups brought suit in U.S. District Court for the District of Arizona, challenging the revised section 10(j) regulations for violating the ESA and APA. The Arizona Game and Fish Commission (AZGFC) intervened in the suit, supporting the revised section 10(j) rules. The federal district court in 2018 during the Trump administration rejected the revised 10(j) management regulations in Center for Biodiversity v. Jewell.

5.3.1 Standard of Review

Judicial review under ESA is governed by the Administrative Procedures Act (APA). The court may set aside agency action where it is found to be arbitrary, capricious, an abuse of discretion, or otherwise not in accordance with the applicable law. The agency's rule is

> arbitrary and capricious if the agency has relied on factors which Congress has not intended it to consider, entirely failed to consider an important aspect of the problem, offered an explanation for its decision that runs counter to the evidence before the agency, or is so implausible that it could not be ascribed to a difference in view or the product of agency expertise.

The arbitrary and capricious standard is "highly deferential," requiring the court to affirm "the agency action if a reasonable basis exists for its decision." The court generally defers to the agency's scientific determinations, which are within the realm of the agency's expertise. Nevertheless, the APA requires the court to engage in a "substantial inquiry" to determine whether the agency acted within the scope of its authority. Although the agency action is presumed to be valid, the court must still conduct a "thorough, probing, in-depth review," of the agency's decision. The court's inquiry into the facts should be "searching and careful."[10]

5.3.2 Population Numbers/Genetics

The FWS determined that a single population of 300 to 325 wolves in the MWEPA with one or two effective migrations per generation would advance conservation. Effective migration involves an individual dispersing between populations, breeding, and producing pups. The FWS did not expect to reach this goal for thirteen years. Nevertheless, the FWS concluded the population objective would ensure the population's persistence and enable it to make further progress toward recovery. The section 10(j) rule was only an interim measure until the revised recovery plan was completed. Finally, the FWS would exercise all management options, particularly translocation, to avoid exceeding the population objective.[11]

The FWS relied on two studies by Drs. Carlos Carroll, Richard Fredrickson, and Phillip Hedrick, which envisioned 750 wolves in three separate interrelated populations of 250 wolves. The FWS then extrapolated from the studies and concluded that one population of 300 wolves with one to two effective migrants joining the population each generation (four years) would eliminate the possibility of long-term extinction risk. The FWS planned to release two packs from the captive population every four years, with the hope of achieving one to two effective migrations per generation. The FWS estimated that between 1998 and 2013, for every 100 wolves released, twenty-one would become effective migrants. If the population grew larger, the number of effective migrations needed would decrease. If genetic variability was not achieved with one to two effective migrations, more wolf pups would be cross-fostered and more captive wolves with pups would be released.[12]

The court in Jewell held the FWS misinterpreted the scientific studies, so failed to utilize the best available science. Prior to the publication of the final rule, Drs. Carroll and Frederickson informed the FWS that it misstated and misinterpreted their scientific findings. The scientists declared in the public comments that the release of two effective migrants per generation from the captive population would not be sufficient to protect the BRWRA population's genetic health and would frustrate recovery. The scientists stressed that their work specifically dealt with the rate of effective migration that would protect the genetic health in a recovered metapopulation consisting of three distinct populations, not in the single isolated genetically compromised BRWRA population. Furthermore, steps to prevent genetic degradation and reduce the high relatedness of the population must be taken at an early stage when the population was small, so they could impact recovery. The court refused to defer to the FWS expertise because the very scientists it had relied on did not support the agency's findings.[13]

The court determined the FWS decision was arbitrary and capricious. The Mexican wolf was facing imminent genetic peril resulting from inbreeding depression. Genetically compromised wolves experience reproductive problems that result in "smaller litter sizes, lower birth weights, and higher rates of pup mortality, as well as lowered disease resistance and other accumulated health problems." The FWS stressed that the Mexican wolf population must be genetically healthy to contribute toward recovery.[14]

The problem with inbreeding dated back to the original McBride lineage, which was founded by only three individuals that were successfully bred. By the mid-1990s, the McBride pups had inbreeding levels similar to offspring from full siblings or a parent-offspring pair. In 1995, the captive breeding program attempting to increase the genetic diversity brought two wolves each from the Aragon and Ghost Ranch lineages, both highly inbred, together with the McBride lineage. The entire Mexican wolf population descended from these seven founders.[15]

Nevertheless, much of the genetic diversity of the founders had been lost. In 2017, the captive population only retained approximately 83 percent of the genetic diversity of its founders, indicating that the problems with inbreeding were not over. The current captive population was only expected to retain 78.5 percent of the genetic diversity of its founders after 100 years, which was lower than the recommended retention of 90 percent of the genetic diversity after 100 years for most captive breeding programs.[16]

The genetic composition of the reintroduced wild population was even worse. The wild population lost 67 percent more of the founders' genetic material than the captive population, and its members were on average almost full siblings.[17] Dr. Frederickson explained that "the reintroduced population is a genetic basket in need of a serious genetic rehabilitation." He warned that failing to address this issue was "irresponsible and also managing for extinction."[18] In addition, Dr. Hedrick pointed out that the current Mexican wolf population only retained genetic material from approximately two of its individual founders. Consequently, the Mexican wolf population "descends from one of the smallest effective founder numbers of any reintroduced endangered species, which portends severe genetic problems."[19]

Compounding the problem was that most of the reintroduced wild population traced their genetic heritage back to a single breeding female wolf. All of the current breeding pairs were producing pups related to this female in the McBride lineage. Dr. Hedrick explained that "this could result in accelerated genetic deterioration because further breeding among descendants of this single wolf could cause lower survival fecundity and greatly reduce the genetic variation of the population."[20] Inbreeding among a small population with low genetic variation would produce an "extinction vortex," which is a self-reinforcing cycle "resulting in decreased fitness and lower survival rates."[21]

The FWS misrepresented the relationship between population size and the extinction risk to the experimental population. The FWS acknowledged that a small isolated Mexican wolf population could not be viable or self-sustaining. Nevertheless, the FWS insisted that its population objective was sufficient to maintain the population and prepare it for the "next phase" of the Mexican wolf's recovery. The court held that the ESA recovery mandate did not focus on "persistence" that only ensured short-term survival. The FWS must be concerned with the long-term viability of the species in the wild, not just its survival. The ESA was designed not only to prevent extinction but to allow recovery so the species could be delisted.[22]

Scientists on the 2012 recovery team asserted that a well-connected metapopulation was important for the recovery of the Mexican wolf. A metapopulation is "a group of populations separated by space but whose members can move between populations or, put another way, several semi-disjunct populations that become viable in the aggregate." The metapopulation should consist of several different Mexican wolf populations consisting of 250 wolves each, separated by space, but whose members could interact. A well-connected metapopulation was "better able to withstand unfavorable demographic rates (e.g., birth rate, fertility rate, life expectancy) and catastrophic environmental events (e.g., wildfire, disease outbreak) than one isolated population." Connectivity between different populations expanded gene distribution, which lessened the adverse impacts from inbreeding. The creation of such a metapopulation would ensure the persistence of the species and their ability to survive a catastrophic event.[23]

The FWS further jeopardized the Mexican wolf's genetic health through the excessive removal of Mexican wolves from the wild. The Mexican wolf was reintroduced as a nonessential experimental species, which allowed the limited taking of individual wolves. However, whenever there was a complaint raised regarding the Mexican wolf (i.e., depredation, human threat), the FWS quickly removed the offending wolf from the wild to placate the opponents of recovery. Between 1998 and 2018, the FWS authorized the removal of 206 Mexican wolves from the reintroduced population.[24] The FWS acknowledged that permanent removals could be equated with wolf mortalities. These removals resulted in the loss of genetically valuable wolves.

5.3.3 Recovery Plan

The FWS argued that the revised section 10(j) regulations were just an interim measure to provide management guidance. There was no direct relationship between these short-term management actions and long-term recovery. Such concerns would be addressed later in the revised recovery plan, which the FWS was in the process of developing.

The court refused to allow the FWS to await the future recovery plan. The court held that the recovery plan did not govern all aspects of recovery, but was a nonbinding statement of intention regarding the agency's long-term conservation goal. The FWS could proceed with recovery under other sections of the ESA, even in the absence of the recovery plan.[25] Moreover, recovery plans were not mandatory, but discretionary. The future recovery plan did not relieve the FWS of section 10(j) requirements.

The court rejected the FWS contention that the final 10(j) regulations were just an interim measure that would not cause any harm to the Mexican wolf in the foreseeable future. It held the FWS misconstrued principles of recovery, which focus on long-term viability, not reliance on future promises. The Mexican wolf experimental population was suffering a genetic decline. The FWS's failure to act was potentially endangering the wolf's genetic health.

The court held that the need for management flexibility did not justify the FWS failure to further long-term recovery. Section 10(j) was enacted to provide long-term flexibility, but did not replace the conservation mandate. On the contrary, management flexibility facilitated the FWS's ability to "conserve and recover endangered species." The FWS was aware that long-term recovery required a consideration of genetic health, so its refusal to acknowledge the dire warnings was an egregious oversight.[26]

5.3.4 Mexican Wolf Experimental Population Area

The section 10(j) regulation abandoned the BRWRA and established the expanded MWEPA, which encompasses all of Arizona and New Mexico south of I-40, totaling 153,871 square miles. The rule prohibited access to suitable wolf habitat north of I-40. Any wolves found outside the MWEPA would be captured and returned to captivity. Such removals would have minimal impact on recovery. From 1998 to 2013, the FWS conducted forty-seven boundary-related removals outside the BRWRA. The FWS explained that any expansion north of I-40 would require coordination with Utah and Colorado. In addition, the MWEPA was not the final Mexican wolf recovery area, which would be determined later in the future recovery plan.[27]

The Coalition of Arizona/New Mexico Counties for Stable Economic Growth intervened in the case, asserting that the I-40 boundary was appropriate. Any expansion above I-40 would imperil the Mexican wolf subspecies. The Mexican wolf was the rarest and most genetically distinct subspecies of all the North American gray wolves. The FWS needed to avoid hybridization with the Northern Rocky Mountain (NRM) wolves, which would undermine the basis for the Mexican wolf's listing as a unique subspecies and ignore the section 10(j) statutory requirement that the experimental population must be "wholly separate from the nonexperimental population." Further, expanding the Mexican wolf's boundary further north would pose a risk to the species. The NRM wolves are larger, run in packs of ten or more, and have a territory of over 500 miles. The NRM wolves would kill Mexican wolves when they meet. The most suitable wolf habitat was actually further south in Mexico.[28]

The court rejected these assertions. Confining the Mexican wolf to the MWEPA was based on politics, not science. The Arizona, Colorado, New Mexico, and Utah governors, along with U.S. Representatives from the area, opposed expansion into the Southern Rocky Mountains (SRM) above I-40. They claimed that the area was not part of the Mexican wolf's historic range and would deprive the states of revenues derived from hunting and recreational activities. Senator Jeff Flake (R. Az.) introduced a bill to prohibit Mexican wolf recovery above I-40, keeping the wolves out of the SRM.[29]

These reservations were dubious. The best available science demonstrated that prohibiting dispersal north of I-40 would jeopardize Mexican wolf recovery. Wolves migrate for various reasons, including competition for prey, breeding, environmental disruptions, and pack dynamics. The benefits of dispersal include greater reproductive success, diminished inbreeding, and less competition for limited prey and habitat.[30]

The FWS confined the wolf in the MWEPA, asserting the area north of 1–40 was not part of the Mexican wolf's historic range. The FWS placed heavy reliance on one article, which asserted the area north of I-40 was not part of the Mexican wolf's historic range.[31] Other scientists criticized the article, which relied on outdated morphological data and was based on Mexican wolf deaths when the population was very low. More recent genetic analysis indicated the Mexican wolf had a broader historic range. Scientists warned that confining the Mexican wolf to the MWEPA would have both short- and long-term adverse genetic consequences.[32]

Prior to extinction, the gray wolf existed across the United States. Strict geographical borders did not exist. Mexican wolves interbred with other gray wolves in the intergradation zone north of I-40, where much suitable habitat exists. The best available science indicated that northern New Mexico, southern Colorado, northern Arizona, and southern Utah contain federal lands and good Mexican wolf habitat. These areas were placed off-limits because of state opposition, not for scientific factors.[33]

Allowing the Mexican wolf population to occupy suitable habitat in the SRM would promote recovery for several reasons: Mexican wolves could coexist with other subspecies and would not be threatened by the larger NRM wolves. In fact, interbreeding with the NRM wolves would introduce valuable and much needed diverse genetic material into the Mexican wolf population, enhancing its adaptability and long-term survival.[34]

Much of the Mexican wolf's southern habitat had become uninhabitable as a result of human activities and climate change, making it insufficient for recovery. Allowing dispersal into the areas north of I-40 would allow the Mexican wolf to adapt to climate change.

Mexican wolves should have been allowed to migrate and perform their role in managing the ecosystem and maintaining biodiversity in the SRM. Congress recognized that "individual species should not be viewed insolation, but must be viewed in terms of their relationship to the ecosystem of which they form a constituent element."[35] The ESA was designed "to return the species to the point where they are viable components of their ecosystems."[36]

5.3.5 Takings

The section 10(j) regulation modified the circumstances under which lethal and nonlethal takings were authorized. This was designed to provide greater management flexibility and improve the chances of success by preventing wolf recovery from interfering with livestock grazing and hunting. If either the AZGFD or the NMGFD determined there were adverse impacts on wild ungulates, the respective agency could request approval from the FWS to remove the wolves from the impacted area.

The FWS asserted that the new take regulations were unlikely to increase Mexican wolf mortality. Since reintroduction there were no reports of significant adverse impacts on ungulates, which was still unlikely with a 300 to 325 wolf population. From 1998 to 2013, the FWS only authorized thirty-six removals, only twelve being lethal, which did not hamper recovery.[37]

The Arizona/New Mexico Counties for Stable Economic Growth intervened, arguing that the revised section 10(j) regulations regarding expanded takings on private land would not endanger the Mexican wolf. The new rule, which allowed the taking of a Mexican wolf caught in the act of killing, was similar to the 1998 regulation. Take permits came with many encumbrances, such as limitations on the number of days allowed for takings and the maximum number of wolves permitted to be taken. A private taking must be done in conjunction with the FWS removals.[38]

The Safari Club International intervened, asserting that the section 10(j) regulation was invalid because it was opposed by state and private stakeholders. Consultation with the states, tribes, and local interests was required by the regulations. In addition, federal regulations must represent "to the maximum extent practicable" a cooperative agreement between state and federal agencies and private landowners.[39]

The court recognized that the expanded take provisions did not provide adequate protection to the valuable Mexican wolves. A take permit is governed by section 10(d), which insists the permit "will not operate to disadvantage of the listed species." The permit must also be consistent with ESA conservation purposes and policies. However, the FWS, when issuing the permits, prioritized the need to coordinate its activities with the stakeholders, particularly hunting and livestock interests, rather than scientific considerations. The expanded take provisions would decrease genetic diversity, which the FWS acknowledged was a major impediment to recovery.

The court held that the section 10(j) regulation did not have to be the product of an agreement between federal, state, and private stakeholders. Congress did not require such a deal. On the contrary, the legislative history demonstrated that Congress encouraged consultation with states and private stakeholders, but the secretary retained final authority and management flexibility. The FWS did consult with the NMGFD and the AZGFD but failed to adopt their suggestions.[40]

The best available science demonstrated that allowing additional taking in response to unacceptable impacts to the ungulate herd would not conserve the Mexican wolf. State agencies were granted too much leeway to define the management goals needed to determine unacceptable impacts. This posed a great threat to the Mexican wolf. The rule would exacerbate the problem of illegal killing, which "is the single greatest source of wolf mortality in the reintroduced population."[41] Liberal taking rules encouraged illegal killing, but did not decrease opposition to wolf recovery.[42] Finally, the new ungulate rule was put in place because Arizona and New Mexico hoped to receive more funds from hunting licenses. This provision was driven by economic, not scientific concerns.

Excessive takings were already jeopardizing Mexican wolf recovery. A study conducted by Dr. Carroll explained that extinction rates were highly dependent upon assumptions about mortality rates. Adult mortality rates greater than the 22.9 percent forecasted in the proposed plan substantially increased the likelihood of extinction. The current mortality rate was over 50 percent. Approximately 112 of the 200 Mexican wolves released since December 31, 2013, had either been killed by humans (86) or permanently removed (16). These were only the documented losses; many more might be undocumented. Larger population sizes were necessary to achieve the optimum extinction rate of 5 percent or less. Even the FWS recognized that the small Mexican wolf population could not sustain such high levels of mortality.[43]

5.3.6 Essential/Nonessential

The section 10(j) regulation retained the Mexican wolf's status as a nonessential experimental population. There are different management strategies for essential and nonessential experimental populations.[44] Essential experimental populations are treated as a threatened species, so are subject to sections 4(d) or 10(j) regulations, which provide more flexibility for their management. They are also subject to the consultation requirement of section 7 of ESA and qualify for the designation of critical habitat. Nonessential experimental populations are treated as a threatened species and are subject to special sections 4(d) or 10(j) regulations. However, solely for the purposes of section 7, nonessential populations are to be treated as a species "proposed to be listed." No critical habitat has to be designated.[45] The FWS asserted that a new essential determination was not required because the 2015 rule simply revised the existing rule.

The court correctly held that releasing the Mexican wolf outside its current range into the MWEPA required a new essential analysis. The Mexican wolf was introduced into the BRWRA in 1998, which constituted its current range. The 2015 regulation expanded the Mexican wolf's current range from 6,854 square miles in the BRWRA to 153,871 square miles in the MWEPA. New releases were allowed in Zones One and Two in the MWEPA, which were outside their current range in the BRWRA. The ESA mandated that an essential determination was necessary if the population was released outside its current range.

The court found the FWS interpretation inconsistent with the plain language of statute. The FWS refused to conduct an essential determination even though all of the statutory requirements were met. In addition, the FWS interpretation frustrated congressional intent that the essential determination must be conducted through informal rulemaking, which was subject to public comment. Finally, the 2015 regulation was not simply a revision of the earlier regulation.

The court determined that the FWS decision to maintain the nonessential designation was not based on the best available science. The FWS evaluation of the Mexican wolf's current status was based on its 1998 status. At that time, eleven wolves were released, with the goal of establishing a self-sustaining population of 100 wolves in the wild, while 150 wolves remained in captivity. The new rule failed to reflect many changes in the Mexican wolf's status: the Mexican wolf was designated as a distinct subspecies; broader releases were permitted in the expanded MWEPA; and dispersals were allowed throughout the MWEPA. Furthermore, the captive population now consisted of 250 aging Mexican wolves, which had lost much of their genetic diversity. The FWS failed to consider the new scientific information generated by these changes.[46]

The FWS was required to perform a new essential determination when authorizing a release of the species outside of its current range. When first introduced into the BRWRA, the Mexican wolf was designated as a nonessential population. This was "necessary to obtain needed State, Tribal, local, and private cooperation," and without it, "the intentional illegal killing of wolves likely would harm the prospects of success." Wolves in the wild were not essential because surplus wolves in captivity could be released. Furthermore, the essential status designation had never been utilized in the past.

A new essential determination was required because the 2015 regulations reclassified the Mexican wolf as a distinct subspecies. The essential determination must take into account the Mexican wolf's changed status. Such determinations must be based on the best available science and biological factors, not political considerations or the need for management flexibility. The 1998 nonessential designation was not based on the best available science. The 2015 rule simply reaffirmed the 1998 rule without any further analysis.[47]

Environmental groups praised the decision. Earthjustice stated: "This ruling offers hope that the Mexican wolf can be pulled back from the brink of extinction before it is too late. The judge made clear that management of the lobo must follow the law and science on Mexican wolf recovery instead of giving in to the political demands of wolf foes." The DOW declared: "The court today reversed what seemed a dim future of the Mexican gray wolf in the Southwest United States. If the lobo is going to have a real chance of recovery, it needs room to roam and protection from unnecessary persecution."[48]

5.4 Final Section 10(j) Regulations

The Biden administration in 2022 published updated section 10(j) regulations in response to the court's decision. The FWS stated: "These proposed revisions are intended to bring the management of the experimental population into alignment with our 2017 Mexican wolf recovery plan."[49] However, the final regulations contradicted several major concerns raised in the Jewell case.

The FWS eliminated the population cap objective of 300 to 325 wolves in the MWEPA, which was consistent with the court's decision in Jewell.[50] The new population objective called for 320 or more Mexican wolves in the MWEPA at the end of eight years. The population must exceed 320 Mexican wolves in each of the last three years, and the population growth must be stable.[51] However, scientists questioned the ability of the two small isolated Mexican wolf populations to avoid extinction. The best available science indicated that a metapopulation, consisting of three interrelated populations of 250 Mexican wolves each, was necessary for Mexican wolf recovery.[52]

The FWS no longer restricted captive wolf releases to one or two per generation. Instead, the FWS planned to annually release a sufficient number of wolves, resulting in twenty-two released wolves surviving to breeding age. The FWS planned to establish an annual schedule for releasing wolves. The goal was to preserve 90 percent of the captive population's genetic diversity in the wild population.[53] This was consistent with the Jewell decision, which recognized that limiting the number of captive wolf releases posed genetic risks.[54]

The FWS, however, continued to rely on twenty-two released wolves surviving to breeding age as its metric of success. However, there was no requirement that the twenty-two released wolves successfully breed to increase the genetic diversity of the population. The record showed that only thirteen wolves released since 2017 reached breeding age. Scientists pointed out that the population abundance was not a good indicator of genetic success.[55]

The FWS only allowed the taking of Mexican wolves pursuant to a permit after annual genetic objectives were met. Takings would only be allowed in the event of unacceptable impacts on the ungulate population. This was consistent with the Jewell decision that prior expanded takings did not provide adequate protection to the Mexican wolf. However, these restrictions were only temporary and contingent on meeting annual genetic objectives, which did not provide adequate protection to other wild-born wolves with valuable genetic mutations.

The FWS did not change the boundaries for Mexican wolf recovery. Mexican wolves were still not permitted to disperse above I-40. Wolves above I-40 could not be taken without a permit.[56] This was not consistent with the Jewell decision, which was critical of the FWS refusal to consider areas north of I-40, where suitable habitat is available for dispersal and recovery. The MWEPA boundary was based on political factors, not the best available science. Precluding the Mexican wolf from occupying suitable habitat did not meet the recovery goals of the ESA.[57] Furthermore, the continued reliance on the second wolf population in Mexico to achieve recovery was problematic.

The FWS retained the Mexican wolf's status as a nonessential experimental population. The FWS concluded that any loss of Mexican wolves in the wild would not be significant. The captive population was robust, so it could easily replace any missing wolves. In addition, Mexican wolves were still present in Mexico. Congress granted the FWS great flexibility regarding nonessential experimental populations and recognized that the FWS must deal with other stakeholders.[58] This was contrary to the court's decision in Jewell that the FWS must conduct a new essential determination because Mexican wolves were being released outside of their current range. The earlier nonessential determination relied on outdated data and was not based on the best available science.[59] Furthermore, the continued reliance on the second wolf population in Mexico was problematic.

The final regulations were criticized by the livestock industry and environmental groups. The livestock industry argued that lifting the population caps threatened ranchers' livelihoods. Environmental groups asserted that the FWS continued to ignore science. The new regulations did not provide for the necessary genetic diversity of the Mexican wolf population; allowed for the taking of Mexican wolves; and failed to abandon the nonessential population designation.[60]

Environmental groups challenged the revised section 10(j) regulations, alleging NEPA, ESA, and APA violations.[61] One of the plaintiffs, the Wild Earth Guardians, stated: "It's outrageous that, over and over, we have to sue the government agency tasked with recovering lobos in order to get them to act according to science, public interest, and law. But if that's what it takes, we will go to the mat for these wolves. They are essential. And they belong."[62] The U.S. District Court for District of New Mexico in CBD v. Haaland rejected many of the earlier findings in the Jewell and upheld the FWS revised section 10(j) regulations. The court, adopting a deferential posture, held that the current genetic and population objectives established by the FWS were valid and would not result in any genetic decline. Restricting the Mexican wolf to the MWEPA and not allowing any migration beyond I-40 was justified. The "nonessential population" designation was consistent with the best available science.[63]

Conclusion

Public law litigation was instrumental in changing the FWS policy, which helped to advance Mexican wolf recovery. Environmental groups petitioned the FWS to change the status of the Mexican wolf in the Obama administration. Facing litigation, the FWS reached a settlement agreement in Center for Biodiversity v. Jewell, which resulted in altering the status of the Mexican wolf to an endangered subspecies. The new status necessitated the change in the section 10(j) management regulations.

The Obama administration proposed revisions to the section 10(j) management regulations in 2015, which were challenged by environmental groups. The federal district court in Center for Biodiversity v. Jewell, adopting a supervisory posture, correctly determined the revised section 10(j) rule violated the ESA and APA. The court held that the population goal of 300 to 325 was not based on the best available science. The FWS's admission that the 10(j) regulation only guaranteed the persistence of the Mexican wolf population was contrary to the ESA conservation mandate. The FWS refusal to consider areas north of I-40, where adequate wolf habitat is available, was questionable. The expanded take provisions did not provide adequate protection to the Mexican wolf. The FWS was required to conduct a new essential determination because Mexican wolves were being released outside their current range. The FWS contention that any problems would be dealt with in the forthcoming recovery plan was rejected. Judicial proponents would support this decision.

The Biden administration proposed revised section 10(j) management regulations in 2022, which were challenged by environmental groups. The federal district court in Center for Biodiversity v. Haaland upheld the revised regulations, which failed to remedy problems identified by the court in Jewell. Judicial skeptics would approve this decision.

The litigation regarding the management of Mexican wolf recovery occurred at the same time that the FWS was revising the recovery plan. There were several overlapping issues in both cases, particularly the use of the best available science, which is dealt with in the next chapter.

Notes

1 77 Federal Register 61375 (October 9, 2012).

2 Steven M. Chambers et al., "An Account of the Taxonomy of North American Wolves From Morphological Genetic Analysis," 77 North American Fauna 1 (October 2012).

3 78 Federal Register 35554 (June 13, 2013). Edward A. Fitzgerald, "Wolf Delisting: Old Wine in New Bottles," 44 Environmental Law Reporter 10413, 10423 (May 2014).

4 *CBD v. Jewell*, No. 1;12-CV-1920 (D.D.C. 2013). 80 Federal Register 2488 (January 16, 2015).

5 Id.

6 80 Federal Register 2512 (January 16, 2015).

7 No. CV-15-00019-TUC-JGZ, No. CV-15-00179-TUC-JGZ, 2018 WL 1586651 (D. Az., 2018), 9–11 (Jewell).

8 Drew Kerr, "Endangered Mexican Gray Wolf Rule Would Hinder Species Recovery, WildEarth Guardians," Plus Media Solutions (November 25, 2014).

"Conservationists Take Aim at Flawed New Rule on Mexican Gray Wolf Management," Center for Biodiversity (January 12, 2015).

9 House Natural Resources Subcommittee on Oversight and Investigation Hearing: The Status of the Federal Government's Management of Wolves; Testimony of Alexandra Sandoval, Director, NMGFD (September 21, 2016).

10 5 U.S.C. section 706(2). *Motor Vehicle Mfts. v. State Farm Mut. Auto. Ins. Co.*, 463 U.S. 29, 43 (1983). Citizens to Preserve Overton Park v. Volpe, 401 U.S. 402, 416-17 (1971).

11 Jewell, 9.

12 Memorandum of Law in Support of Plaintiffs Motion for Summary Judgment, *WildEarth Guardians v. Ashe*, No. CV-15-00285-TUC-JGZ (February 23, 2016), 22–7. (WEG Memo)

13 Jewell, 10–17.

14 Id., 14.

15 78 Federal Register 35 664, 35, 704–5 (June 13, 2013).

16 Plaintiffs Memorandum in Support of Motion for Interim Injunctive Relief (July 20, 2018), 2.

17 Complaint for Declaratory Relief (Complaint), *CBD v. Jewell.*

18 Jewell, 12.

19 Philip Hedrick, "Genetics & Recovery Goals for Mexican Wolves," 206 Biological Conservation 210 (2017).

20 Plaintiffs Memorandum in Support of Motion for Interim Injunctive Relief (July 20, 2018).

21 Jewell, 12.

22 Id., 14–15.

23 Hedrick, 210.

24 Amended Complaint for Declaratory and Injunctive Relief, *CBD v. Jewell*, 23–4.

25 Id.

26 Jewell, 14–17.

27 Federal Combined Memo in Support of Cross Motion for Summary Judgement, 24–5.

28 Plaintiffs Coalition of Arizona and New Mexico Counties for Economic Growth, et al.'s Memorandum In Opposition to Center for Biodiversity, et al.'s Memorandum In Support of Motions for Summary Judgment, 3–6.

29 Brady McCombs, "Utah Balks at Being Part of Recovery Zone for Mexican Wolf," Albuquerque Journal (December 6, 2015). "Chaffetz Leads Western States Coalition in Fighting the Introduction of the Mexican Wolf In Utah," Congressional Documents & Publications (December 11, 2015). "Senator Flake Introduced Bill the Supplant Science in Endangered Mexican Wolf Recovery, Suppress Population," CBD (February 16, 2017).

30 Western Environmental Law Ctr, WildEarth Guardians, New Mexico Wilderness Alliance, Memorandum of Law in Support of Plaintiffs Motion for Summary Judgment (February 23, 2016) 27–31. (WELC Memo).
31 J. R. Heffelfinger et al., "Clarifying Historical Range to Aid Recovery of the Mexican Wolf," 81 Journal of Wildlife Management 766 (2017).
32 WELC Memo, 27–31.
33 WELC, Public Comments Processing, Attn. FWS-R2-ES-2020-0007 (June 15, 2020), 19–21.
34 Earthjustice, Public Comments Processing, Attn: FWS-R2-ES-2017-0036 (August 20, 2017), 3–10.
35 House of Representatives Conference Report. No. 97-835, at 30 (1982).
36 House of Representatives Report, No. 95-1625, n.5 (1978)
37 Combined Memorandum in Support of Cross-Motion for Summary Judgment and Opposition to Motion for Summary Judgment in No. 4:15-cv-00019-JGZ at 24 (April 25, 2016) (Federal Combined Memo).
38 Plaintiffs, Memorandum in Opposition to CBD et al.'s Memorandum in Support of Motion for Summary Judgment, 11–12.
39 Federal Combined Memo, 28 n.12.
40 Jewell, 13–17.
41 WELC Memo, 31–4.
42 Francisco J. Santiago-Avila et al., "Liberalizing the Killing of Endangered Wolves was Associated with More Disappearances of Collared Individuals in Wisconsin, USA," 10 (1) Scientific Reports 13891 (August 17, 2020).
43 Memorandum in Law in Support of Plaintiffs Motion for Summary Judgment, 13–14, 32–4 (Ashe Memo).
44 "Essential" means the experimental population's loss "would be likely to appreciably reduce the likelihood of the survival of the species in the wild. 50 C.F.R. 17.80 (b).
45 16 U.S.C. 1533 (a)(3)(A). 16 U.S.C. 1539 (j)(2)(C)(i), (ii).
46 Jewell, 19–21.
47 Ashe Memo, 9–15.
48 "Court rejects flawed Mexican gray wolf rule,": Center for Biodiversity (April 2, 2018).
49 86 Federal Register 59953 (October 29, 2021) (draft regulations). 87 Federal Register 39348 (July 1, 2022).
50 Jewell, 14, 20–1.
51 87 Federal Register, 39349–50.
52 Carlos Carroll et al., "Biological and Sociopolitical Sources of Uncertainty in Population Viability Analysis for Endangered Species Recovery Planning," Scientific Reports (July 12, 2019).

53 87 Federal Register, 39350.

54 Jewell, 23 n.13.

55 Lindsey Botts, "New Rules Would Lift Limits on Mexican Wolf, but Activists Say Changes Fall Short," Arizona Central (October 30, 2021).

56 87 Federal Register, 39350–3 Jewell, 28–9.

57 Jewell, 13–14, 23 n.13. See also, "Mexican Wolves See Improvement by Imperfect Management Rule," WildEarth Guardians (October 29, 2021).

58 87 Federal Register, 39351–3.

59 Jewell, 19.

60 "FWS Faces Criticism over Finalized Mexican Wolf Rule (10j)," New Mexico Politics Report (July 5, 2022).

61 U.S. District Court for the District of Arizona, Complaint for Declaratory and Injunctive Relief, *CBD v. Haaland*, Case 4:22-cv-00303-JAS (July 13, 2022). "FWS Sued for Failing in Mexican Gray Wolf Recovery," National Parks Traveler (July 13, 2022).

62 Adrian Hedden, "Lobo Population Cap Lifted, but Environmentalists Say New Rules Don't Go Far Enough," Carlsbad Current Argus (July 7, 2022).

63 No. CV-22-00303-TUC-SHR (March 31, 2025). AZGFD, Court rules in favor of federal and state agencies on Mexican wolf management rule (April 21, 2025).

6

Mexican Wolf Recovery Plans

The development of the Mexican wolf recovery plans from the Obama through the Biden administrations was controversial. The resulting litigation, like the earlier section 10(j) management case, focused on the use of best available science in the decision-making process. The earliest recovery plan was developed during the Reagan administration in 1982, which the FWS acknowledged was problematic. This chapter analyzes the litigation in Defenders of Wildlife v. Zinke dealing with the first revised Mexican wolf recovery plan developed in the Obama administration; the litigation in Center for Biodiversity v. Haaland regarding site-specific actions and objective management criteria in the first revised plan in the Biden administration; and the litigation in Wild Earth Guardians v. Haaland upholding the second revised wolf management plan in the Biden administration.

6.1 Recovery Plan

Section 4(f) of the ESA directs the secretary of interior (secretary) to "develop and implement [recovery] plans . . . for the conservation and survival" of species listed as endangered.[1] Recovery is "the process by which the decline of a threatened species or endangered species is arrested or reversed and threats to its survival are neutralized, so that its long-term survival in nature can be ensured."[2] Recovery is more than preventing extinction. Recovery ensures that the protected species exists in sufficient numbers and is sufficiently well-distributed, so that ESA protections are no longer needed. Recovery is based on the viability of the species in the wild, not in captivity. Recovery planning "is the 'umbrella' that eventually guides all . . . activities [under the act] and promotes a species' conservation and eventual delisting."[3]

Recovery planning began as an administrative action. In 1978, Congress mandated that recovery plans be implemented. Recovery was linked to delisting in the 1986 regulations, which declared that "recovery means improvement in the status of listed species to point at which listing is no longer appropriate under the criteria set out in section 4(a)(1) of the act."[4]

The 1988 ESA amendments required the secretary, when developing and implementing a recovery plan, to incorporate "to the maximum extent practicable" in each plan: (1) a description of such site-specific management actions necessary to achieve the conservation and survival of the species; (2) objective, measurable criteria which, when met, would allow the species to be removed from the list; and (3) estimates of the time required and the cost to carry out these goals. Furthermore, the secretary must "provide public notice and an opportunity for public review and comment on such plan," then "consider all relevant information presented prior to approval of the plan."[5]

6.2 History of the Mexican Wolf Recovery Plans

The FWS completed the first Mexican wolf recovery plan in 1982 during the Reagan administration. The principal objectives of the recovery plan were to maintain the captive breeding program of 250 wolves and reestablish a "viable, self-sustaining population of at least 100 wolves in the middle to high elevations of a 5000 square mile area within Mexican wolf's historic range by releasing captive wolves into Southwest United States." The FWS noted that "experts have rated the recovery of Mexican wolf subspecies as the highest priority of all such programs." There was a great deal of criticism of the 1982 recovery plan. The plan acknowledged that illegal killing had to be addressed, but contained no criteria to assess the threat and no preventative measures. The FWS acknowledged that the 1982 recovery plan was "far from complete" and violated the ESA.[6]

Mexican wolves were introduced into the Blue Range Wolf Recovery Area (BRWRA) in 1998. The first three-year review of the Mexican wolf recovery program was completed in June 2001 during the George W. Bush administration. The report noted that the Mexican wolf population consisted of approximately thirty-five of the ninety-nine wolves released since March 1998. Some wolves had been shot illegally, hit by vehicles, captured and returned to captivity, or died of natural causes. Significantly, the wolves were functioning as a population: killing natural prey; forming pairs; and reproducing in the wild.[7] During the first three years, there were fourteen instances of livestock damage attributed to the wolves. Ranchers were compensated by the Defenders of Wildlife (DOW) compensation fund, which had been established in 1987 and expanded in 1995 to cover losses in the southwest.[8]

The report recommended that the recovery plan be revised because the small target population of 100 wolves was not viable for the Mexican wolf's long-term recovery. Mexican wolves should be directly released into Gila National Forest in New Mexico, which contained 77 percent of the BRWRA, 700,000 acres of roadless areas and negligible livestock. Wolves should be allowed to establish territories outside the BRWRA, where adequate habitat was available. Wolves leaving BRWRA should not be recaptured because as the population grows, dispersals would occur. Mexican wolves should be allowed to establish dens on private land, if they did not cause problems. Reintroduction into White Sands Wolf Recovery Area should be reconsidered. Livestock operators on public lands should be held responsible for carcass management and disposal. Human-caused wolf mortality must be decreased.[9]

The Adaptive Management Oversight Committee (AMOC) completed the five-year program review in January 2006 during the Bush administration. Several of the following recommendations were controversial. States and tribes should be able to use nonlethal measures to harass wolves. Wolves could be taken once the population exceeded 125, and grazing fees on public land occupied by wolves should be reduced. Environmental groups complained that the federal government ignored the critical recommendations of the earlier 2001 three-year assessment, which was the only independent assessment ever conducted. They noted that "hostility toward science is undermining the wolf recovery program."[10]

The FWS 2010 five-year assessment during the Obama administration warned that the recovery program was "at risk of failure" because of inadequate regulations, illegal shootings, and inbreeding. The FWS stressed that the failure to develop an up-to-date recovery plan was hindering recovery.[11] The Arizona Game and Fish Commission (AZGFC) in 2011 voted to oppose the release of any new wolves from captivity into the state until a new recovery plan was completed. The AZGFC later agreed to allow replacement wolves to be introduced into the state.

The FWS in 2010 put together a team of eminent scientists to develop a revised recovery plan. They produced the 2012 draft recovery plan, which called for the establishment of a metapopulation consisting of three interconnected populations, each with at least 250 wolves, in the BRWRA, which encompassed southern Utah/northern Arizona and southern Colorado/northern New Mexico. Delisting could occur after natural dispersal between the three populations produced at least one genetically effective migrant every generation, averaged over the eight-year period. The annual rate of human-caused mortality averaged over the eight-year period had to be less than 20 percent. In addition, a post-delisting monitoring plan and state regulatory programs must be in place.[12]

The draft proposal generated controversy. The FWS decided not to proceed with the 2012 proposal ostensibly because the status of the Mexican wolf and section 10(j) management regulations were being revised. Both actions were being challenged by environmental groups. Critics, however, asserted that the FWS abandoned the effort in the face of political pressure from the affected states.[13]

In response, the Public Employees for Environmental Responsibility (PEER) filed a complaint with the U.S. Department of Interior's Office of Integrity, stating that "the scientific integrity of the work of the Science and Planning Subgroups of the Mexican Wolf recovery team has been significantly and intentionally compromised by political interference of the FWS, as well as by specified state 'Partners.'" The PEER further alleged that the FWS's abandonment of the 2012 draft recovery plan violated President Obama's Scientific Integrity Policy issued in January 2011, which stated: "When scientific or technical information is considered in decision making, the information will be as robust, of the highest quality, and the result of the most rigorous scientific processes as can be achieved within the available decision time frame." In the PEER's view, politics interfered with the recovery process, particularly Utah's and Colorado's objections. The PEER accused the Obama administration of "pre-negotiating the science to accommodate its political partners" and preventing the recovery team from proceeding with the draft Mexican wolf recovery plan. Representative Grijalva (D. N.M.), the ranking Democratic minority member on the House Natural Resources Committee, demanded an investigation, arguing that "attempts to change scientific findings because of political preferences should not be part of the process."[14]

Environmental groups in November 2014 filed suit in the U.S. District Court for the District of Arizona, alleging the FWS violated section 4(f) by failing to prepare a new recovery plan. Colorado, Utah, and the New Mexico Game and Fish Department (NMGFD) intervened in the suit. Arizona filed a separate suit. At the same time, the AZGFC and NMGFC continued to prevent the release of adult wolves from captivity into their states until a new recovery plan was implemented.[15]

There was an unsuccessful effort in Congress demanding a new recovery plan. Senators Jeff Flake (R. Az) and John McCain (R. Az) introduced the Mexican Wolf Recovery Plan Act, which required the FWS to draft a new Mexican wolf recovery plan with state and local government participation. If the FWS failed to cooperate with Arizona and New Mexico, the states would assume management of the recovery process.[16]

While awaiting the court's decision, the FWS began a closed-door process with personnel from Arizona, Colorado, New Mexico, Utah, and Mexico to develop a revised recovery plan. Scientists responsible for the 2012 draft recovery plan were not invited to participate in the process, which did not focus on utilizing the best available science to promote Mexican wolf recovery.[17]

The Obama administration negotiated a settlement agreement in October 2016, which required the FWS to complete the final recovery plan by November 30, 2017. The FWS would conduct an independent peer review of the plan with the participation of New Mexico, Colorado, Utah, and Arizona. The FWS would submit status reports every six months and assume all costs. The NMGFD and Colorado refused to join the agreement because they objected to the final plan deadline, but both agreed to voluntarily dismiss their claims.[18]

6.3 2017 Recovery Plan

The FWS released the long-awaited recovery plan for the Mexican wolf in June 2017 during the Trump administration. The final plan called for a population of 320 Mexican wolves in the United States and 200 Mexican wolves in Mexico over an eight-year period. In addition, the US population average over the eight-year period must be equal to or greater than 320 wolves. The population must exceed 320 wolves in each of the last three years of the eight-year period. The annual population growth must be stable or increasing. Preservation of the genetic diversity of the captive population would be ensured by twenty-two Mexican wolves surviving to breeding age. In Mexico, the wolf population average over the eight-year period must be greater than or equal to 200 wolves and must exceed 200 wolves in each of the last three years of the eight-year period. The annual population growth must be stable or increasing. The genetic diversity of the captive population would be ensured by the survival of thirty-seven wolves surviving until breeding age. And, states and tribes in the United States and Mexico must have regulatory mechanisms in place to prohibit or regulate human-caused mortality of Mexican wolves.[19]

Environmental groups criticized the plan. The Center for Biodiversity (CBD) noted: "This is not a recovery plan, it's a blueprint for disaster for the Mexican wolf. By limiting their habitat and stripping protections too soon, this plan ignores the science and ensures the Mexican wolf never reach sufficient numbers to be secure." The Defenders of Wildlife (DOW) declared: "Once again, politics trumps science. The final recovery plan fails the Mexican wolf with inbreeding, dangerously low population, insufficient range and intense trapping and shooting. Mexican gray wolves are not receiving the science-based plan they desperately need to survive."[20]

The NMGFD approved the Mexican wolf recovery plan, which recognized that approximately 90 percent of the subspecies historical habitat exists in Mexico. However, the New Mexico Cattlegrowers Association was "disappointed that the program is expanding, but this was to be expected." Senator Jeff Flake characterized the plan as "another federal regulatory nightmare for ranchers and Arizona rural communities," which failed to consider the needs of the local stakeholders, including the Arizona Cattle Growers Association, and allowed the Mexican wolf population to expand indefinitely.[21]

Efforts to curtail Mexican wolf recovery in Congress continued. Senator Flake introduced a bill that called for the FWS to delist the Mexican wolf and return wolf management back to the states. House Republicans introduced a provision to the 2018 Department of Interior's Appropriation bill that halted and stopped the funding for all wolf recovery efforts across the country, including Mexican wolf recovery. The provision was not adopted, but another section requiring the FWS to examine the Mexican wolf's genetics to assess its status as a subspecies was enacted.[22]

6.4 Center for Biological Diversity v. Zinke

Environmental groups brought suit challenging the plan in the U.S. District Court for District of Arizona during the Trump administration. Section 11 of the ESA allows a person to commence suit "against Secretary to perform any act or duty under [section 4] of this title which is not discretionary with the Secretary." The provision was designed "to encourage enforcement [of the ESA] by a so called private attorney general out of a recognition that the overall subject matter of this legislation is the environment in which it is common to think all persons have an interest."[23] The CBD alleged the recovery plan violated the ESA because the FWS failed to base its population and genetic goals on the best available science, jeopardizing the wolf's genetic diversity. In addition, the FWS disregarded the best available science regarding suitable Mexican wolf habitat in the United States. The FWS relied on recovery efforts in Mexico, where there was a lack of suitable habitat and weak management.

The court, adopting a deferential posture, rejected the CBD's arguments. The court concluded that the CBD's allegations were disagreements regarding the FWS policy determinations, which fell within the agency's discretion, so were not reviewable under section 11. The court also held that the FWS was not required to utilize the best available science in the recovery plan. Section 4(f) did not contain any textual reference mandating the use of the best available science, which was only required for listing and delisting determinations.[24]

The court's deference to the FWS's decisions was consistent with prior precedents, which held that recovery plans were not binding documents and not subject to detailed substantive review. However, scholars have questioned judicial temerity regarding the review of recovery plans. Scholars have attempted to link recovery plans to the ESA "duty to conserve" in section 7(a)(1) of the ESA in order to encourage greater judicial scrutiny of recovery plans.[25] For example, Professor Oliver A. Houck stated: "The case can be made that, since section (7)(a)(1) of the ESA requires all federal agencies to 'conserve' endangered wildlife species, and that since the ESA defines 'conservation' in terms of species recovery, recovery plan elements are powerful limits, if not mandates, for agency action."[26] Professor Federico Cheever noted it was inconceivable "that actions by federal agencies, which actively frustrate recovery, do not contravene the agencies' duty to utilize their authority in furtherance of the purposes of [ESA] by carrying out programs for the conservation of endangered species and threatened species."[27] Professor Daniel Rohlf declared: "Recovery plans in many ways possess the ideal characteristics to act as triggers for agency's duty to conserve listed species." They are prepared by experts and "often identify which federal agencies are responsible for carrying out specific recovery tasks." Restricting agency actions to the duties designated in the recovery plan "would free the courts from sticky problems of attempting to interpret the scope of ESA conservation mandate on case-by-case basis."[28]

The court failed to acknowledge that the FWS discretion regarding the recovery plan was not unlimited. The ESA required the secretary to perform several nondiscretionary functions regarding the recovery plan. The ESA mandated that the secretary "shall develop and implement plans . . . for the conservation and survival of the endangered and threatened species." Recovery plans that did not provide for species conservation and recovery should be reviewable. Furthermore, the secretary "shall" to "maximum extent practicable" address three statutory factors—site-specific management actions, objective, measurable criteria, and time and cost estimates.[29] These were nondiscretionary functions, which should have been subject to judicial review under section 11.

There were other problems with the court's decision. The court failed to recognize the importance of public input into the recovery plan. The secretary was required to seek public input when developing the recovery plan under section 4(f) of the ESA. This provision was an amendment offered by Senator James McClure (R. Id.), who explained that "this amendment will require the Secretary to solicit comments and additional information for consideration from local communities prior to federal approval of the new recovery plan, or before approved revisions to existing plans." The amendment was not intended to make the secretary "responsible for gathering the information." Instead, it sought to "allow those most directly affected by a proposal to provide additional information to the Secretary that might otherwise be overlooked." The final congressional report stated that section 4(f) did "not necessitate a rule making procedure," but did require the secretary to "consider the public comments before approving the plan."[30]

Congress developed procedures for making changes to the recovery plan in response to changed conditions. These changes were also subject to public notice, which "ensures that affected parties have an opportunity to participate in and influence agency decision making at an early stage, when the agency was more likely to give real considerations to alternative ideas." If the elements of the recovery plan could be changed at the secretary's discretion without following the prescribed process, the public input mandated under section 4(f)(1) became superfluous.[31] Notably, the U.S. District Court for the Southern District of California, stressing the importance of the recovery plan, "respectfully disagreed with the cases minimizing the important of recovery plans" and asserted that "the ESA clearly requires the FWS to follow through with the measures identified in the recovery plans."[32]

The Zinke court also held that the recovery plan did not have to be based on best available science, which only applied to listing and delisting decisions.[33] This was contrary to several earlier court decisions,[34] the legislative history, and the FWS's own policy. The legislative history confirmed that the FWS was required to utilize the best available science in the recovery plan. Senator McClure offered three amendments to the ESA in 1988, which required recovery plans to contain (1) an annual report of expenditures, (2) provisions for public review, and (3) an analysis of the economic impacts. The first two passed, but Senators John Chafee (D. R.I.) and George Mitchell (D. Me.) opposed the inclusion of economic considerations. Senator Chafee declared that recovery plan decisions must "be based solely on biology." Economic considerations could not be factored into the process. Senator Mitchell concurred, declaring that recovery must be based exclusively "on the biological status of the species, not on how much it will cost to restore its numbers or whose economic interest it may affect." Senator Mitchell stated: "It is essential to the act's integrity that recovery goals for the number of individuals or populations needed to ensure a species existence be based solely on the best available scientific evidence."[35]

The FWS also recognized that the recovery plan must be based on the best available science. The FWS Service Recovery Planning Guidance (SRPG) stated that recovery plan criteria must be "grounded in good science and defensible." The SRPG also noted that "recovery plans delineate such reasonable actions as may be necessary, based upon best science and commercial data available, for the conservation and survival of the listed species." In addition, the SRPG peer review guidance "requires the Service to make biological decisions based upon the best science and commercial data available. These decisions involve listing, reclassification, and delisting, critical habitat designations, and recovery plan implementation."[36] The court should not have accorded any deference to the FWS litigation position that was inconsistent with agency practice.[37]

Scholars support the utilization of best available science in recovery plans. There are many benefits of the best available science requirement. It demonstrates that the agency's decision is based on science, not politics. There will be greater public trust in the decision because it does not benefit any particular interest. The judiciary is more likely to defer to an agency's action that is based on science. Finally, the nature of the decision-making process will be improved. Professor Holly Doremus noted: "It seems unlikely that Congress intended to condone the production of recovery plans . . . without reference to the best available science." Even though Congress did not expressly include such a requirement "it goes without saying that agencies making decisions with a high scientific content must consult the best available science."[38]

Professor Dan Goble stated: "The recovery plan is a statement of the best available scientific data on the conservation management actions needed to protect and recover a listed species." Agency decisions must be predicated upon precisely such data. Courts have employed the science in recovery plans to establish basic facts about the biology of species and their status. In other cases, the courts have turned to recovery plans to provide an independent source of scientific information. The plans will generally be evidence of the best available science since they are not developed in the context of an individual dispute.[39]

Since the recovery plan was the prelude to delisting, the best available science requirement should have been required. Otherwise, a flawed recovery plan that did not provide for the conservation of the Mexican wolf would be implemented, frustrating the goals of the ESA to delist the species. It was better to act early to address problems, rather than later to solve problems. There were 163 Mexican wolves in the MWEPA in 2019.

6.4.1 Site-Specific Management Actions/Objective, Measurable Criteria: Illegal Killing

The court did, however, conclude that the FWS failed to include site-specific management actions or objective, measurable criteria to deal with the illegal killing of Mexican wolves. The FWS identified the problem of human-caused mortality, but did not provide any suggestions on how to address the problem or explain why it was not practicable to do so.[40]

The court's decision was supported by text of the ESA. Explicit congressional commands leave no room for agency discretion. The ESA states that recovery plans "shall" include objective measurable criteria. The use of "shall" is an imperative denoting a specific obligation. The requirement "to maximum extent practicable" indicates strong congressional preference that the agency fulfill its obligations to the fullest extent possible. The goal of the objective measurable criteria is to remove the species from the list of threatened or endangered species. The objective measurable criteria must address each of five statutory delisting factors and measure whether threats to the species have been eliminated.[41]

The court's decision was consistent with legislative history, which demonstrated the importance of recovery plans. Congress was concerned with flaws in prior recovery plans. The Senate Report on the 1988 ESA amendments noted that recovery plans must contain "site-specific management actions to achieve recovery [and] criteria by which to judge success of the plan." This would "ensure that plans are as explicit as possible in describing the steps to be taken in the recovery of a species."[42]

Nevertheless, scientists continued to criticize the 2017 plan. Sixty scientists urged the FWS to adopt an "entirely new approach to the management and recovery of the Mexican wolf—an approach based on science, acknowledgement of past shortcomings, humaneness, and a precautionary approach to the management of this genetically unique and genetically depleted regional subspecies."[43] Dr. Carroll noted: "Recovery plan goals based on politics rather than science slow Mexican wolf recovery by allowing the agency to forego opportunities to establish new populations in suitable habitats and to underestimate the number of wolves needed to be released from captivity into the wild population to improve genetic health." He suggested that "the FWS needed to go back to the drawing board and come up with a science-based plan that would truly put the Mexican wolf on the path to recovery."[44]

Environmental groups appealed the case to the Ninth Circuit Court of Appeals, alleging the 2017 plan failed to provide for the conservation and survival of species and was not based on the best available science, as the law required.

6.5 Center for Biological Diversity v. Haaland

The court in CBD v. Zinke determined that the recovery plan failed to include site-specific management actions that address human-caused mortality. In response, the FWS amended the recovery plan. The litigation continued in the Biden administration. There were 196 Mexican wolves in the MWEPA in 2021.

6.5.1 Site-Specific Management Actions

In the follow-up case, the FWS alleged the amended plan provided numerous and sufficient site-specific management actions in the recovery action table at the end of the plan. The FWS carried out site-specific management actions through its implementation strategy, which articulated more detailed site-specific, near-term actions necessary to carry out provisions in the recovery plan. The implementation strategy was not a statutory mandate, so it was not subject to public review. The FWS acknowledged that the implementation strategy was not a substitute for the recovery plan, but argued that the strategy should be considered as further evidence that the plan adequately addressed human-caused mortality in its site-specific actions.

The U.S. District Court for the District of Arizona in CBD v. Haaland, following its earlier holding, rejected the FWS assertions. The court held that when a specific threat was recognized in the recovery plan, the FWS must recommend corrective action or explain why this was not practicable. The FWS did neither and failed to show why the management actions identified in the implementation strategy were not included in plan. The FWS had a clear duty to fulfill the statutory mandate. The plan was remanded to the FWS. The FWS was ordered to produce a draft plan within six months, which must include site-specific management actions to reduce the number of wolves illegally killed.[45]

Environmental groups praised the court's ruling. The CBD stated: "The path to recovery for the Mexican wolf has been hampered by widespread poaching for far too long. Now that the FWS has to take this issue seriously, we hope these wolves will stand a better chance of survival." The DOW stated: "This court ruling recognizes the urgent need for the FWS to develop a recovery plan that addresses a significant threat to the Mexican wolf, poaching by people." Nevertheless, environmental groups remained skeptical. The CBD noted that the FWS still maintained a mentality of "persecution" toward the wolf. The latest management plan was "woefully inadequate for rescuing the Mexican wolf."[46]

6.5.2 Objective Measurable Criteria

There were other problems with the court's decision. The objective measurable criteria must address the five ESA delisting factors in the recovery plan, specifically assessing whether the threats originally leading to the listing of the species have been remedied in ways that would permit biological recovery of the listed species. The court held that the objective measurable criteria in the revised plan addressing human-caused mortality fell within the bounds of agency discretion, so were not subject to judicial review. Furthermore, the court found the FWS objective measurable criteria regarding the protection of genetic diversity, the adequacy of regulatory measures, and the availability of suitable habitat were reasonable.[47] These conclusions were problematic.

6.5.2.1 Human-Caused Mortality

The FWS acknowledged that human-caused mortality posed a continual threat. The FWS addressed this threat through population abundance, which required the US population to average at least 320 wolves and the Mexican population to average at least 200 wolves over eight years. Each population must be stable or growing. The FWS asserted that population abundance "ensures that the population is resilient and threats of human-caused mortality have been ameliorated." The FWS noted that "mortality rates will need to be sufficiently low to achieve recovery criteria because they are a primary indicator of wolf population trajectory." The plan utilized a Population Viability Analysis (PVA) to synthesize biological information on factors affecting the demographic and genetic status of the species, and to predict the influence of these factors on population viability and endangerment. The most important parameter affecting extinction risk was adult mortality. The PVA concluded that mean adult mortality rates had to be "less than 25%, combined with mean sub-adult mortality rates less than 33% and mean pup mortality . . . less than 13% for there to be an increasing population that meets the population abundance recovery criteria." The FWS considered these mortality rates to be acceptable and appropriate measures for assessing whether human-caused mortality has been ameliorated.[48]

The FWS analysis was dubious. The FWS acknowledged that human-caused mortality constituted the greatest source of wolf deaths during recovery. Between 1998 and 2020, 74 percent of the Mexican wolf deaths (119 of 2016) were caused by humans. Fourteen Mexican wolves were killed in 2020, eight were killed illegally and six died after being hit by vehicles. Since reintroduction, only ten individuals pled guilty to poaching.[49]

The peer reviewers concluded that the recovery plan did not address illegal killing. Peer review ensures that the FWS decisions are based on sound scientific data, assumptions, and analysis. Peer reviewers evaluate the scientific information utilized in the decision-making process; gather information that disputes the official position; require the government to document its positions; and ensure the use of the best available science in the decision-making process. Peer reviewers, who are scientists, are shielded from political pressure, so have a different perspective than government scientists.[50] Dr. Stetz, a peer reviewer, noted the draft recovery plan lacked any objective measurable criteria addressing illegal killing, whereas other recovery plans typically utilized age and class/sex-specific percentages of abundance estimates to clearly express mortality limits. Dr. Carroll concurred that the recovery plan did not contain any criteria addressing the threat of human-caused mortality.[51]

Population targets were inadequate to ensure the conservation and survival of the Mexican wolf. Dr. Carroll pointed out that the FWS population recovery targets did not account for the high level of mortality experienced, but assumed an unnaturally low rate of mortality. The FWS overestimated the proportion of females in the breeding pool each year and assumed wolves would be released into the wild at the forecasted rate. The FWS vested control over releases in the states that had in the past worked to oppose and delay releases. As a result, the FWS vastly underestimated how many wolves would be needed to ensure population viability.[52]

The population goals were not based on the best available science, but were the product of negotiations between the FWS and state agencies based on socioeconomic concerns. Dr. Carroll asserted that the PVA was manipulated to achieve the desired result. The FWS had to meet politically acceptable wolf population numbers, while countering the probability of extinction. However, "any recovery plan that relies on misrepresentation of scientific data faces long odds in effectively achieving recovery." A population based on politics, not science, would not survive.[53]

The FWS's own recovery planning guidance was critical of utilizing population abundance to address human-caused mortality. The planning guidance warned that "merely increasing a species' numbers, range and abundance does not ensure its long-term health or sustainability; only by alleviating threats can lasting recovery be achieved." The planning guidance concluded that "evaluating a species for potential reclassification or delisting requires an explicit analysis of threats under the five listing factors in addition to the evaluation of population or demographic parameters."[54] Conservation measures might increase the Mexican wolf population, while the threat of human-caused mortality remained unabated. For example, the FWS provided supplementary feeding for 70 percent of denning Mexican wolves to reduce wolf-livestock conflict. This might result in lower mortality. The Mexican wolf population might increase, but the threat posed by human-caused mortality would not be alleviated.[55]

6.5.2.2 Inadequate Regulatory Mechanisms

Federal courts have insisted that "regulatory mechanisms" essential to the species viability must consist of specific legal mandates with "some method of enforcing compliance." In addition, state plans must be "sufficiently certain and effective to alleviate a threat of endangerment . . . after delisting."[56] The FWS delisting criterion required states and tribes to establish regulatory mechanisms that ensured at least 320 Mexican wolves would be maintained post-delisting. The court found this criterion to be objective and measurable because the Mexican wolf population had to reach 320 prior to delisting.[57]

Reliance on states and tribes to have adequate regulations programs in place was problematic. This criterion was vague and reliance would fail to address the continued killing of Mexican wolves even while they remain protected by the ESA. Mortality thresholds that would signal threats to the Mexican wolf and regulatory mechanisms that would be effective to adequately diminish illegal killing were not identified. The promise to put in place future regulatory actions was not an adequate substitute for objective measurable criteria to stop illegal killing. In addition, the FWS acknowledged that the killing of wolves in the United States would likely increase after delisting because state penalties are weaker than federal penalties.[58]

The transfer of authority over endangered and threatened species to state governments has been viewed with great skepticism.[59] Several scholars cautioned that "devolution of federal authority and responsibility over threatened and endangered species to states is likely to undermine conservation and recovery efforts, lead to a greater number of species becoming imperiled, and result in fewer species recovered."[60] Other scholars pointed out that the FWS "has a spotty . . . track record when it comes to assessing the adequacy of regulatory mechanisms and deploying them as a substitute for ESA protections."[61]

6.5.2.3 Genetic Diversity

The FWS acknowledged that the loss of genetic diversity posed a threat to Mexican wolf recovery and survival. The FWS concluded that genetic diversity would be achieved when twenty-two Mexican wolves released from the captive population survived to breeding age in the wild population. This meant that a pup must live for at least two years after being released. In addition, much of the genetic diversity of the captive population must be retained in the wild population. The FWS concluded that the preservation of 90 percent of the genetic diversity of the captive population in the wild population would be achieved. The court considered this to be reasonable and refused to consider any alternative methods.[62]

The court's decision was questionable. The FWS acknowledged that inbreeding depression could affect the probability of a breeding pair of wolves successfully producing a litter. Much of the genetic diversity of the founders had been lost. The captive population only retained 83 percent of the founders' genetic diversity, which was lower than the recommended 90 percent retention rate for most captive breeding programs. The wild reintroduced population faced even greater genetic peril. Only 75.48 percent of the founders' population had been retained in the US wild population. In the Mexican population, only 73.88 percent of the founders' genetic diversity remained.[63] In addition, the current Mexican wolf population only possessed genetic material from approximately two founders. Members of the Mexican wolf population were, on average, related to each other as full siblings. This high relatedness posed the risk of inbreeding depression, which would jeopardize recovery.[64]

The FWS underestimated the impact of inbreeding depression. Dr. Carroll and others noted that the PVA and the recovery plan failed to adequately consider genetic threats. Inbreeding depression was not incorporated into the model when predicting the probability of litter size; therefore, the recovery criteria were "inadequate to address the genetic threats that arise due to small population size."[65]

The FWS estimated that seventy wolves from the captive population would have to be released into the wild to have twenty-two released wolves survive until breeding age.[66] This was questionable because there was no requirement that the twenty-two released wolves breed and increase the genetic diversity of the population. Furthermore, linking the gene diversity goal of the wild population to the already genetically compromised captive population was likely to exacerbate genetic threats.[67]

Excessive removal of Mexican wolves from the wild also proved to be an impediment to achieving genetic diversity. The FWS authorized or carried out the removal of 206 Mexican wolves from the reintroduced population between 1998 and 2019. Many of these removals were due to depredation. The ten-year (2011–20) average depredation rate was 58 deaths per 100 cows. However, ranchers were not required to take steps to minimize depredation. Often, there was no postmortem on the carcasses to prove wolves were responsible for the killing. These removals included wolves that could have improved the genetic diversity of the wild population.[68]

The genetic threats were aggravated because the recovery plan failed to provide for connectivity between the United States and Mexican wolf populations. The existing border barriers and the proposed border wall hindered connectivity and will be addressed in the next chapter.[69] The recovery plan envisioned two separate populations in Mexico and the United States. The FWS concluded that there was limited opportunity for even a low level of connectivity between the MWEPA and the northern Sierra Madre Occidental (SMOC) wolf populations. The FWS assumed a nonbreeding wolf between one and four years of age only had a 0.175 percent annual probability of dispersing between the MWEPA and the SMOC. Wolves arriving after migrating from their source population only had a 37.5 percent survival rate, so the expected average successful dispersal rate would be approximately one individual dispersing from the MWEPA population to the SMOC population every twelve to sixteen years (three to four generations).[70]

The FWS determined the lack of any connectivity between the United States and Mexican populations would have no long-term impact on the survival of the Mexican wolf. The growth potential and survival rate in the two separate populations would be sufficient to ensure survival of the species and compensate for the lack of connectivity. The FWS acknowledged that several Mexican wolves had migrated from Mexico but did not remain in the United States.[71] Nevertheless, the FWS recognized that the lack of connectivity between the two populations presented a problem: the loss of genetic diversity. The FWS stressed that other methods might be necessary to compensate for the loss of connectivity, such as more translocations and greater releases. This would be essential for the wolf population in SMOC, which was smaller and dispersal possibilities were lower.[72]

The lack of connectivity demonstrated that the recovery plan was not based on the best available science. Scientists stressed that connectivity between the two populations was crucial. Larger populations with genetic diversity have a greater chance of survival because they are able to withstand stochastic events. Populations with genetic diversity are able to avoid problems of inbreeding. Small populations, like the Mexican wolf, were particularly vulnerable to these problems. Scientists asserted that one effective migration per generation would be adequate to avoid any genetic deterioration. If this could not be achieved, a lower level between 0.5 and 0.1 effective migrations per generation would be sufficient to provide for adaptive connectivity. This would maintain the historic genetic structure in the region.[73]

6.5.2.4 *Suitable Habitat*

The FWS concluded that the loss of habitat did not constitute a threat to Mexican wolf recovery. There was sufficient habitat in the MWEPA and Mexico to ensure recovery. The court held that the conclusion was reasonable, so the FWS was not required to develop additional habitat criteria.[74] The court's decision was problematic. The FWS failed to recognize the importance of suitable habitat north of I-40. The same court in CBD v. Jewell was skeptical of the section 10(j) regulation prohibiting Mexican wolf dispersal above I-40 into the SRM.[75]

Mexican wolves need large blocks of land with a sparse human population and low livestock density. The FWS recognized that the most suitable wolf habitat occurs on public lands, particularly in national forests. There are large blocks of public land in northern Arizona, southern Utah, the Grand Canyon, the San Juan National Forest in southern Colorado, and the nearby Carson National Forest in northern New Mexico, which would provide suitable habitat. The SRM, which extend from south-central Wyoming to northern New Mexico, have been described as "the mother lode for wolves." Early FWS studies showed the SRM area could support 1,100 wolves.[76]

The absence of wolves in the SRM region represented a significant gap in the taxon. Since the region is equidistant from the Northern Rocky Mountains and the MWEPA, the establishment of the SRM wolf population would create "a spatially segregated population of wolves that extended from the Arctic to Mexico." Dr. David Mech, a noted wolf expert, declared "[SRM] restoration could connect the entire North American wolf population from Minnesota, Wisconsin, Michigan through Canada and Alaska, down the Rocky Mountains into Mexico. It would be difficult to overestimate the biological and conservation value of this achievement."[77]

The recovery plan's reliance on Mexico was also problematic because Mexico's ability to help was questionable. Wolf reintroduction in Mexico began in 2011 with the release of five Mexican wolves (three females, two males). Four were killed within a month. The first wild birth occurred in 2014. From 2012 to 2016, forty-one Mexican wolves were released into the SMOC, eighteen died within a year, many as a result of illegal killing. In 2017, there were thirty-five Mexican wolves in the northern SMOC in the states of Sonora and Chihuahua. However, many of the surviving wolves had to be fed by humans to discourage them from roaming into other riskier areas. Supplementary feeding is very expensive, which can be an impediment to additional releases.

Land ownership in Mexico presents additional problems for wolf recovery. Mexico has three types of land ownership: federal, private, and communal. The federal government does not own much land. The few naturally protected areas in Mexico lack management plans, permit extractive industries, and have inholdings. Community property, called "Ejidos", is generally degraded and managed for multiple use, not for sustainability. In addition, Mexico has higher livestock densities than the United States, which has resulted in a great deal of wolf/livestock conflicts. Illegal human-caused mortality also poses significant danger south of the border. Overall, Mexico lacks suitable habitat, an adequate prey base, and a weak legal framework to support Mexican wolf recovery.[78]

6.6 The 2022 Recovery Plan

The Biden administration, responding to the court's decision, released an updated recovery plan in September 2022, which retained all elements of the 2017 recovery plan and included site-specific management actions to address illegal killings. The plan articulated several options for downlisting and delisting the Mexican wolf. It included provisions that promised to conduct education and outreach to local communities, increase law enforcement presence in mortality "hot spots" to deter illegal killing, investigate wolf mortalities, coordinate with local law enforcement, install underpasses or warning signs in key areas to help wolves cross roads, require ranchers to use proven techniques to protect cattle, such as electric fencing and range riders, and pursue efforts to increase genetic diversity, including more cross-fostering. Mexican wolves were still not allowed to disperse above I-40.[79] There were 242 wolves present in the MWEPA in 2022.

6.7 Wild Earth Guardians v. Haaland, Center for Biodiversity v. Haaland

Environmental groups filed an appeal in the Center for Biodiversity v. Zinke case in the Ninth Circuit, arguing the 2022 recovery plan did not preclude their appeal because the plan was virtually identical to the 2017 plan. The Ninth Circuit declared the case moot in 2023. The court held that the issuance of the 2022 recovery plan superseded the 2017 recovery plan. Substantial changes in the

2022 plan precluded the court from determining whether the 2017 plan provided for the conservation and survival of the Mexican wolf.[80] The court was mistaken. Many of the flaws identified in the 2017 recovery plan were still present in the 2022 plan, which only added the site-specific management actions regarding illegal killing. The remainder of the 2022 plan simply mirrored the 2017 plan, so the issues raised on appeal remained unresolved. Case law demonstrates that issues on appeal that were not resolved by current administrative action, which incorporate elements of the original action, remain subject to judicial review.[81] There were 257 wolves in the MWEPA by the end of 2023.

Conclusion

Public law litigation was not successful in altering the Mexican wolf recovery plans. The federal district court in Center for Biodiversity v. Zinke upheld the 2017 recovery plan. The court held that the citizen suit provisions in the ESA could not be invoked because the recovery plan was discretionary agency action. The court, adopting a deferential role, failed to recognize that the FWS's obligation to develop a recovery plan that provided for conservation and survival of the species was not a discretionary action. In addition, the court mistakenly determined that the recovery plan did not have to be based on best available science. Nevertheless, the court, adopting a supervisory posture, invalidated part of the 2017 recovery plan because it did not contain site-specific management actions and objective measurable criteria to address the illegal killing of Mexican wolves.

In the follow-up case, Center for Biodiversity v. Haaland, the federal district court again determined that the FWS revisions to the 2017 plan failed to contain site-specific management action to address illegal killings. However, the court, reverting to a deferential position, determined that the objective management criteria regarding human-caused mortality, the adequacy of regulatory measures, the protection of genetic diversity, and the availability of suitable habitat were reasonable. These conclusions were problematic. Ironically, these cases were argued before Judge Jennifer Guerin Zipps, who had invalidated the 2015 revised section 10(j) management regulations in CBD v. Jewell.

The FWS issued a revised recovery plan in 2022. Environmental groups filed an appeal in the Ninth Circuit challenging the court's 2017 Zinke decision. The Ninth Circuit held that the 2022 plan invalidated the 2017 plan, so the case was moot. The Ninth Circuit decision was dubious because the same problems presented in the 2017 plan regarding nature of the recovery plan and employment of the best available science were not resolved in the 2022 recovery plan.

The courts granted too much deference to the FWS regarding the recovery plans and failed to carry out the ESA mandates. These cases demonstrate the weakness in the recovery plans, which are only considered to be a preliminary procedural document. According to the courts, the FWS must only address the three statutory requirements in the recovery plan. Rigorous review of the substance of the recovery plan is precluded because the methods to achieve the conservation and survival of the Mexican wolf are considered to be policy decisions left up to the agency's discretion. The FWS is not required to employ the best available science because the recovery plan is only a preliminary document that is subject to change. This undermines the ESA mandate to provide for the conservation and recovery of the Mexican wolf so that statutory protections are no longer needed. Judicial skeptics would applaud these decisions, which involved judicial deference and did not interfere with federal policy.

Notes

1 16 U.S.C. 1533 (f)(1).

2 Federico Cheever, "Recovery Planning, the Courts, and the Endangered Species Act," 16 Natural Resources & Environment 106 (2001–2).

3 60 Federal Register 11,768, 11,770 (March 2, 1995)

4 Cheever, 107–8. See also, Jason M. Patlis, "Recovery, Conservation, and Survival under the ESA: Recovering Species, Conserving Resources, and Saving the Law," 17 Public Land & Resources Law Review 55, 68–75 (1996).

5 16 U.S.C. 1533 (f)(1), (4), (5).

6 CBD et al. Plaintiffs Memorandum of Points and Authorities in Support of Motion for Summary Judgment, No. 4:18-cv-00047-JGZ at 8 (April 17, 2020) (CBD Plaintiffs Memo).

7 Paul C. Paquet, John A. Vucetich, Michael K. Phillips, and Leah M. Vucetich, *Mexican Wolf Recovery: Three Year Program Review and Assessment* (U.S. Fish and Wildlife Service, June 2001).

8 By the end of 2003, the fund had spent $27,288 in the southwest.

9 Tania Soussan, "Gray Wolf Findings Applauded," Albuquerque Journal (May 9, 2001).

10 "Wolves: Expanded Boundaries, More Controls for SW Program," Land Letter (December 22, 2005).

11 FWS, *Mexican Wolf Conservation Assessment 2010*, 14. April Reese, "Southwest Reintroduction Program Flawed—FWS," Land Letter (May 13, 2010).

12 For a full discussion of the 2012 proposal, David R. Parsons, "Comments on the Draft Recovery Program," Rewilding Institute (August 24, 2017), 1–7.

13 Matt Robinson, "Mexican Gray Wolf Management Plan for Gila National Forest, Western New Mexico, Dropped by Feds," Las Cruces Sun-News (New Mexico) (February 20, 2013).

14 April Reese, "Endangered Species: Another Controversy in Mexican Wolf Program as Watchdog Files Integrity Complaint," Environment & Energy News (June 20, 2012).

15 Earthjustice, "Lawsuit Fights 38 Years of Delay in Recovering Southwest Mexican Gray Wolf," Plus Media (November 20, 2014). "NM Game Commission Rejects Fed Releases of Mexican Wolves into Gila National Forest," Targeted News Service (September 29, 2015).

16 Press Release, "Sens. Flake, McCain Call for Revised Mexican Gray Wolf Recovery Plan" (April 29, 2016).

17 Parsons, 12.

18 *Defenders of Wildlife v. Jewell*, 2016 U.S. Dist. LEXIS 145122 (D. Az., October 18, 2016) (DOW). "Court Guarantees Mexican Gray Wolves Recovery Plan," Targeted News Service (October 18, 2016).

19 82 Federal Register 29918 (July 1, 2017). U.S. Fish and Wildlife Service, The Mexican Wolf Recovery Plan, First Revision 2017, 18–20.

20 "Trump Administration Finalizes Deeply Flawed Mexican Wolf Recovery Plan," Center for Biological Diversity (November 29, 2017).

21 "New Mexico: Approval of Fed Wolf Plan Could Signal Thawing Relations," Greenwire (January 4, 2018). "FWS Plan Calls for Release of 12 Newborn Pups," Santa Fe New Mexican (December 5, 2017).

22 "Endangered Species: Senate Bill Targets MW Protections," Greenwire (January 9, 2018). U.S. Official News, "House Republicans Expand War on Wolves in Latest Budget Bill."

23 16 U.S.C. 1540 (g). *Bennett v. Spear*, 520 U.S. 154, 165 (1997).

24 399 F. Supp. 3d 940, 946–9 (D. Az. 2019).

25 16 U.S.C. 1536.

26 Oliver A. Houck, "The Endangered Species Act and Its Implementation by the U.S. Departments of Interior and Commerce," 64 University of Colorado Law Review 277, 350 (1993).

27 Federico Cheever, "The Road to Recovery, A New Way of Thinking About the Endangered Species Act," 23 Ecology Law 1, 59–60 (1996).

28 Daniel J. Rohlf, *The Endangered Species Act: A Guide to Its Protections and Implementation* (Stanford Environmental Law Society, 1989), 98. See also, J. B.

Ruhl, "Section 7(a) of the 'New' Endangered Species Act," 25 Environmental Law 1107, 1151–2 (1995).

29 16 U.S.C. 1533 (f)(1).

30 House of Representative Conference Report, 100–928 at 21.

31 *Friends of Blackwater v. Salazar*, 591 F. 3d 428, 440–8 (2012) (Rogers, dissent).

32 *Southwest Center for Biological Diversity v. Bartel*, 470 F. Supp. 2d 1118, 1137 n.16 (S.D. Cal. 2006).

33 399 F. Supp. 3d, 949.

34 *Fund for Animals v. Babbitt*, 903 F. Supp. 96, 110 n.4 (D.D.C 1995), 114. *Defenders of Wildlife v. Babbitt*, 130 F. Supp. 2d 121 (D.D.C. 2001).

35 134 Congressional Record 19273 (1988).

36 Appellants Joint Opening Brief, *Wild Earth Guardians v. Haaland*, Ninth Circuit (2023), 37–8.

37 *Defenders of Wildlife v. Norton*, 258, F.3d 1136, 1145 n.11 (9th Cir. 2001).

38 Holly Doremus, "The Purpose and Effects, and Future of ESA's Best Available Science Mandate," 34 Environmental Law 397, 407, 418–19 (2004).

39 Dale D. Goble, *Recovery, in Endangered Species Act, Law, Policy, and Perspectives*, 2d ed., eds. Donald C. Baur and Williams Robert Irvin (American Bar Association, 2010), 87–90.

40 399 F. Supp. 3d, 949–50.

41 *Fund for Animals v. Babbitt*. 108–11

42 Senate Report No. 100–240, 4, 9–10 (1987).

43 "Competing Interests Leave MW Recovery Eat Crossroads," Arizona Republic (December 31, 2019).

44 Carlos Carroll et al., "Biological and Sociopolitical Sources of Uncertainty in Population Viability Analysis for Endangered Species Recovery Planning," Scientific Reports (July 12, 2019). "Study: Politics Harming Recovery of Mexican Wolf," Center for Biodiversity (July 12, 2019).

45 *Center for Biodiversity v. Haaland*, 562 F. Supp. 3d 68, 80–62, 70 (D. Az. 2021).

46 Peter Aleshire, "Judge Orders FWS to Come Up with Plan to Decease Wolf Poaching," Payson Round-Up (November 26, 2021).

47 *Center for Biodiversity v. Haaland*, 562 F. Supp. 3d at 83–7.

48 Id.

49 Adrian Hedden, "Human Killing of Endangered Mexican Wolf Addressed in Federal Plan," Carlsbad Current-Argus (April 18, 2022).

50 59 Federal Register 34271 (July 1, 1994). Jeremy Bruskotter et al., "Expert Judgment and Uncertainty Regarding Protection of Imperiled Species," 31 Conservation Biology 657 (June 2017).

51 CBD Plaintiffs Memo, 11

52 Amended Complaint for Declaratory Relief, 20.

53 Western Environmental Law Center, Public Comment-Endangered and Threatened Wildlife and Plants: Revision to the Nonessential Experimental Population of the Mexican Wolf (Canis lupus bailey) at 6 n.29, 17–19.

54 Center for Biodiversity et al., Plaintiffs Memorandum in Opposition to Federal Defendants' Motion for Summary Judgment, *Center for Biodiversity v. Bernhardt*, No. 4: 18-cv-00047-JGZ, 17 (Plaintiffs' Memo)

55 Complaint for Declaratory and Injunctive Relief, *CBD v. Zinke*, No. 4: 18-cv-00047-BGM (January 30, 2018) 20. (Complaint Zinke)

56 *Oregon Natural Resource Council v. Daley*, 6 F. Supp. 2d 1139, 1155 (D. Or. 1998). *Defenders of Wildlife v. Zinke*, 849 F.3d 1077, 1083 (D.C. Cir. 2017).

57 562 F. Supp. 3d, 84–5.

58 CBD Plaintiffs Memo, 19–21.

59 Holly Doremus, "Delisting Engendered Species: An Aspirational Goal, Not a Realistic Expectation," 30 Environmental Law Reporter 10434, 10446 (2000). Federico Cheever, The Rhetoric of Delisting Under the Endangered Species Act," 31 Environmental Law Reporter 11302 (2001).

60 Alejandro E. Camacho et al., "Assessing State Laws and Resources for Endangered Species Protection," 47 Environmental Law Reporter 10837, 10838,10843 (2017).

61 Sandra B. Zellmer et al., "Species Conservation & Recovery Through Adequate Regulatory Mechanisms," 44 Harvard Law Review 367, 368 (2020).

62 562 F. Supp. 3d, 85–6.

63 Complaint Zinke, 15.

64 Earth Justice, "Public Comment Processing" (August 29, 2017), 32–3.

65 Complaint Zinke, 20–1.

66 Federal Defendants Brief in Support of Combined Cross-Motion for Summary Judgment, *Center for Biodiversity v. Bernhardt*, No. 4: 18-cv-00047-JGZ (May 22, 2020), 28.

67 Complaint Zinke, 19–20

68 Complaint for Declaratory and Injunctive Relief, *Center for Biodiversity v. Haaland*, Case 4:22-cv-00303-JAS (July 12. 2022), 23–4.

69 "Border Wall Poses Problems for Endangered Mexican Wolf," New Mexico Political Report (October 30, 2019).

70 Phillip S. Miller, *Population Viability Analysis for Mexican Wolf: Integrating Wild and Captive Populations in Metapopulation Risk Assessment Model for Recovery Planning*, Report prepared for the U.S. Fish and Wildlife Service (2017), 47.

71 U.S. Fish and Wildlife Service, Mexican Wolf Recovery Plan, Second Revision (September 2022), 8

72 Miller, 49.

73 Carlos Carroll, Richard J. Fredrickson, and Robert C. Lacy, "Developing Metapopulation Connectivity Criteria from Genetic Habitat Data to Recover the Endangered Mexican Wolf," 28 Conservation Biology 76–86 (2013).

74 562 F. Supp. 3d, 87.

75 *Center for Biodiversity v. Jewell*, 33–4.

76 Michael Robinson, Dave Parsons, and Rob Edward, "South from Yellowstone, What Remains to Be Done," International Wolf (Spring 2006), 8, 9.

77 Michael K. Phillips, Rob Edward, and Tina Arapkiles, "Restoration of the Gray Wolf to the Southern Rocky Mountains," in Nina Fascione, Aimee Delach, and Martin E. Smith (eds.), *People and Predators: From Conflict to Coexistence* (Island Press 2004), 240, 243–6.

78 Earth Justice, "Public Comments Processing," 14–18.

79 Susan Montoya Bryan, "Proposed Revision of Mexican Wolf Management Plan Draws Fire," Associated Press (May 13, 2022).

80 *WildEarth Guardians et al. v. Haaland*, Case No. 22-15029; *Center for Biological Diversity et al. v. Haaland*, Case No. 22-15056 (9th Cir. 2023). "FWS Sued for Failing in Mexican Gray Wolf Recovery," National Parks Traveler (July 13, 2022).

81 *350 Montana v. Haaland*, 50 F.4th 1254, 1263–4 (9th Cir. 2022).

7

Mexican Wolves Confront the Border Wall

Mexican wolf recovery has been adversely impacted by border barriers in Arizona, New Mexico, and Texas. Border barriers fragment Mexican wolf habitat and territory on the border. They stop Mexican wolf migration and dispersion between the two countries. This prevents the genetic exchange between the populations, which is necessary for the species' health. Electric lighting on the barriers affects the behavior and nocturnal movements of the Mexican wolf. The noise pollution generates stress that can cause harmful metabolic and hormonal behavior problems. The barriers hamper collaborative efforts with Mexico regarding Mexican wolf recovery.[1]

This chapter reviews the history of border barrier construction and its effect on Mexican wolf migration. The public law litigation during the George W. Bush and Donald J. Trump administrations regarding the secretary of homeland security's (SHS) utilization of the waiver authority to facilitate wall construction and President Trump's transfer of funds to construct his border wall are analyzed. Subsequent events regarding the border wall during the Trump and Biden administrations are discussed.

7.1 History and Resulting Litigation Before the Trump Administration

Numerous fences were proposed and constructed along the southwest border from 1924 to 1955. The Immigration and Naturalization Service (INS) in 1977–8 approved funding for the construction of border fences in El Paso, Texas, San Ysidro, California, and San Luis, Arizona. These fences replaced older, dilapidated fences and diverted illegal immigrants away from cities into the desert where they could easily be apprehended. In October 1978, the INS announced that the new twelve-foot-high fences would be capped with barbed wire. These fences became known as the "Tortilla Curtain."[2]

7.1.1 William J. Clinton Administration

The legal framework governing border wall construction began in the Clinton administration.[3] Congress strengthened border security by enacting the Illegal Immigration Reform and Immigrant Responsibility Act (IIRIRA) of 1996, which was attached as a rider on an appropriation bill. Policymaking through appropriation riders undermines deliberative lawmaking for many reasons: First, legislation through appropriation riders does not receive adequate consideration. Second, the enactment of legislation through appropriation riders alters the balance of power in Congress. Third, appropriation riders enact legislation that cannot get through the front door of the conventional legislative process, but by the back door of the appropriation process. Fourth, appropriation riders interfere with the president's ability to veto legislation. Fifth, policymaking through appropriation riders enhances interest group influence in Congress.[4]

The IIRIRA was the first legislative enactment that specifically mandated the construction of physical barriers and roads. Section 102(a) granted the US attorney general (AG) authority to construct border barriers. Section 102(b) directed the AG to install additional physical barriers and roads in areas of high illegal entry into the United States along 14 miles of the southwest border in the San Diego, California sector.[5] This area defined the scope of the AG's waiver authority under section 102(c), which was limited to the National Environmental Policy Act (NEPA) and Endangered Species Act (ESA). The Clinton administration built 24 miles of border fencing in southern California.

7.1.2 The George W. Bush Administration

Congress intensified its effort to improve border security after the 9/11 tragedy. In 2002, the newly created Department of Homeland Security (DHS) assumed responsibility for border security. In response, the INS was dissolved and its duties were transferred to the Bureau of Customs and Border Protection (CBP). Binational efforts to protect and manage the environment along the southwest border were not a priority during the Bush administration.

Legislative efforts to expand the SHS waiver authority continued. The Real ID Act (RIDA) of 2005 expanded the SHS's waiver authority under section 102(c) to cover all barriers constructed under the IIRIRA. The SHS was allowed to waive "all legal requirements" the secretary in her sole discretion "determines necessary to ensure expeditious construction of the barriers and roads." The RIDA also limited judicial review. Federal district courts could only review constitutional claims, which have to be filed within sixty days of the action by the SHS. The appellate jurisdiction of the Circuit Courts of Appeal was removed. District court decisions regarding the use of the waiver could only be reviewed by the Supreme Court regarding constitutional issues.[6]

The demand for additional border fencing continued. The Secure Fences Act of 2006 (SFA) amended section 102(b). The SHS was directed to "provide for 2 layers of reinforced fencing, [and] the installation of additional physical barriers, roads, lighting, cameras, and sensors" along 698 miles in five areas in five specific segments along the US–Mexico border across California, Arizona, New Mexico, and Texas.[7] The CBP estimated that because of topographical features the SFA actually mandated fencing across 850 miles. These areas circumscribed the extent of the waiver authority under the RIDA. Prominent Democrats, who later opposed President Trump's border wall, supported the SFA, including Senator Barbara Boxer (D. Cal.), Senator Barack Obama (D. Ill.), Senator Joe Biden (D. Del.), Senator Dianne Feinstein (D. Cal.), and Representative Chuck Schumer (D. N.Y.)

Many border barriers were constructed during the second term of the Bush administration. As of December 31, 2008, the DHS had constructed 578 miles of fencing.[8] Secretary Chertoff issued five waivers, covering thirty-five laws, which authorized the construction of border barriers on 645 miles along the southwest border. Most of these waivers affected territory that could be utilized by Mexican wolves to cross the border. Environmental groups challenged SHS Chertoff's invocation of the waiver authority. However, federal district courts supported the SHS waivers in the areas designated by Congress.

7.1.2.1 Sierra Club v. Ashcroft

On September 13, 2005, Secretary Chertoff waived eight statutes to permit the expeditious construction of 14 miles of border fence in San Diego. The Sierra Club brought suit, alleging that the waiver violated the nondelegation doctrine and unlawfully constrained the district court's jurisdiction. In 2005, the U.S. District Court for the Southern District of California in Sierra Club v. Ashcroft rejected all the allegations and dismissed the case. Nevertheless, the court noted that the SHS waiver authority was limited to the San Diego fencing authorized by the IIRIRA.[9]

7.1.2.2 *Defenders of Wildlife v. Chertoff*

On October 26, 2007, Secretary Chertoff waived twenty laws to allow the construction of a little over 7 miles of border fencing through the San Pedro Riparian National Conservation Area (SPRNCA) in Arizona, a potential Mexican wolf migration route. The Bureau of Land Management (BLM) granted the DHS perpetual right of way. The BLM completed an Environmental Assessment (EA), but did not conduct an Environmental Impact Statement (EIS) pursuant to the NEPA. The Defenders of Wildlife (DOW) brought suit, alleging that the EA was inadequate and the NEPA required a full EIS. In addition, the BLM's action violated the Arizona-Idaho Conservation Act of 1988, which instructed the BLM to manage the SPRNCA "in a manner that conserves, protects, and enhances the riparian area" and to "only allow such uses of the conservation area" that further the purposes of the Act.[10] The U.S. District Court for the District of Columbia in DOW v. Chertoff upheld the SHS waivers. The court held the nondelegation rule was not violated because the waiver was limited to the "expeditious completion" of the border fences authorized by the IIRIRA in areas of high illegal entry.[11]

The RIDA allowed limited appeal exclusively to the Supreme Court regarding constitutional questions. The plaintiffs appealed to the Supreme Court, arguing that section 102(c) was unconstitutional because it violated the nondelegation doctrine and granted the SHS legislative power to "void any federal law, free of any review of his determinations." However, the Court denied the petition for certiorari on June 23, 2008.[12]

7.1.2.3 *Save Our Heritage v. Chertoff*

On January 19, 2007, the secretary waived nine laws to allow construction on 42.4 miles of border barriers and roads in San Diego and the Goldwater Air Force Range in Arizona, a potential Mexican wolf border crossing. The Save Our Heritage Organization (SOHO) brought suit, seeking an injunction to halt the construction. The SOHO alleged the SHS action violated numerous statutes and that the waiver of these statutory requirements was unconstitutional.

The U.S. District Court for the Southern District of California in 2008 upheld the secretary's decision. The court held that the general purpose of section 102(a) provided the SHS with "general authority to construct border barriers." Furthermore, the SHS waiver authority under section 102(c) did not violate the nondelegation doctrine.[13]

7.1.2.4 El Paso v. Chertoff

Secretary Chertoff waived thirty-seven federal laws to allow the construction of 560 miles of border fencing in four states along the border, which also would impede Mexican wolf migration. On the same day, Secretary Chertoff waived twenty-seven laws to allow the construction of 21.4 miles of border fencing in Hidalgo County, Texas. The suit brought by the County of El Paso challenging these waivers was dismissed by the U.S. District Court for the District of West Texas.[14]

The plaintiffs again petitioned the Supreme Court for a writ of certiorari, arguing that the unlimited discretion granted to the SHS to waive all statutes interfering with border wall construction violated the nondelegation doctrine. The Court rejected the petition on June 15, 2009.[15]

7.1.3 The Barack H. Obama Administration

President Obama was not a proponent of border fencing. During a presidential debate in 2008, candidate Obama stated: "I think that the key is to consult with local communities, whether it's on the commercial interest or the environmental stakes of creating any kind of barrier."[16] SHS Janet Napolitano, while governor of Arizona, often voiced her skepticism of border fencing, stating: "You show me a 50-foot wall, and I'll show you a 51-foot ladder at the border."[17] Nevertheless, Secretary Napolitano decided to allow border fence projects already contracted under the Bush administration to continue, but disregarded Chertoff's waiver for a segment of vehicle barrier on the Tohono O'odham Nation in southwest Arizona. President Obama in 2011 declared that fencing along the US–Mexican border was "now basically complete."[18]

Later in the year, the Obama administration had to deal with wolves from Mexico crossing the border into the United States. Mexico began reintroducing wolves into the wild close to the US border in October 2011. The Arizona Game and Fish Department (AZGFD) and environmental groups applauded the move as "extraordinary good news," pointing out the San Madre Occidental Mountain range possessed good wolf habitat. The FWS did not share this enthusiasm because the migrating wolves were protected by the ESA. The FWS in November 2011 granted itself a permit to trap and indefinitely incarcerate any wolves entering Arizona and New Mexico from Mexico. The Center for Biodiversity (CBD) brought suit, challenging the FWS permit in 2013. Several weeks later, the FWS rescinded the permit. In the resulting 2013 settlement agreement, the FWS rescinded the regulation, acknowledging its lack of authority to issue the permit.[19] Mexican wolves could continue to cross the border.

As of May 2015, the DHS had installed 353 miles of primary pedestrian fencing, 300 miles of vehicle fencing (total 653 miles), 36 miles of secondary fencing behind the primary fencing, and 14 miles of tertiary pedestrian fencing behind the secondary fence. The CBP had identified a total of 653 miles of border as appropriate for fencing and barriers.[20] This circumscribed the extent of the SHS waiver authority.

7.2 President Trump's Border Wall

The Trump administration faced several problems regarding the border wall. Congress refused to fund the wall, so President Trump transferred funds to build his wall. States and environmental groups challenged the transfer of funds and the SHS utilization of the waiver authority to construct his border wall.

7.2.1 Border Wall Funding

President Trump envisioned 700 to 900 miles of a see-through wall across the US–Mexico border. He promised that Mexico would fund the wall, but Mexico refused to pay for the wall. On January 25, 2017, President Trump issued Executive Order 13767, the "Border Security and Immigration Enforcement Improvements," which instructed the DHS to construct a wall along much of the 2000-mile-long US–Mexico border.[21]

President Trump initially sought funding for his border wall through the normal appropriation process, but Congress refused. The issue came to a head at the end of 2018. Congress passed a bipartisan continuing resolution, which kept the remaining federal agencies funded through 2019, but included no funding for the border wall. After criticism from conservative commentators, President Trump refused to sign the bill and executed a partial shutdown of the federal government. During the shutdown, he announced that if negotiations regarding his border wall were not successful, he would "call a national emergency and build [the wall] very quickly."[22]

The Democrats took over control of the House of Representatives in 2019. After the 35-day partial shutdown was temporarily halted, a bipartisan committee was created to reach a budget compromise. The January 2019 compromise provided $1.375 billion for 55 miles of border fencing in the Rio Grande Valley in Texas.

President Trump signed the agreement, but issued a proclamation declaring a national emergency on the southern border. The national emergency declaration was designed to get around congressional spending restrictions and transfer military funds to pay for the border wall. President Trump stated that "he went through Congress," but was "not happy" with the $1.375 billion appropriated for his wall. As he explained, "I didn't need to do this. But I'd rather do it much faster . . . that's all."[23]

President Trump then announced plans to divert $6.1 billion in federal funds for border wall construction. The administration identified three funding sources: $3.6 billion from military construction projects; $2.5 billion from other military accounts under counterdrug authorities; and the remaining $601 million from the Treasury Forfeiture Fund.

The House and Senate for the first time passed resolutions opposing the national emergency declaration. President Trump vetoed the joint resolution. The House failed to override the veto.

The secretary of defense (SOD) began to divert $1 billion in military funding for the border projects in New Mexico and Arizona. The Sierra Club and Southern Border Communities Coalition (SBCC) brought suits seeking an injunction to stop the transfer of funds for violating section 8005.[24] At the same time, another suit brought by California and New Mexico sought an injunction to halt the transfer of $1.5 billion for four border projects in California and Arizona. In May and June 2019, the U.S. District Court for the District of Northern California in the SBCC and the California cases granted injunctions stopping the transfer of funds under section 8005.[25]

The federal government petitioned the Ninth Circuit to grant a stay of the district court injunction. The Ninth Circuit in July 2019 rejected the federal government's request. The court held that the federal government was unlikely to succeed on the merits of the case. The transfer of funds did not meet the requirements of section 8005. Furthermore, the Trump administration's action violated congressional authority over appropriations.[26]

The federal government appealed to the U.S. Supreme Court. On July 26, 2019, the Supreme Court reversed the lower courts and granted a stay of the district court's preliminary injunction. The stay would remain in effect until the federal government's appeals were exhausted. In addition, the Court raised questions regarding the plaintiffs' standing. Justices Ginsburg, Sotomayor, and Kagan voted to deny the petition. Justice Breyer argued for a partial stay that would allow the government to finalize the contracts, but not to begin construction. The Court's decision allowed the federal government to spend $1.6 billion for the construction of bollard fencing on 130 miles along the US–Mexican border in Arizona, California, and New Mexico, hampering Mexican wolf migration.[27]

Professor Vladeck noted that this case represents the Supreme Court's use of the Shadow Docket, such cases not decided on the merits but dealing with procedural matters in ongoing litigation brought to the Court, which result in "unsigned procedural orders shaping and structuring how the justices process and ultimately resolve each of the cases before them." The Shadow Docket was not a problem in the past, but in recent years the Court has been using it to intervene prematurely in controversial policy disputes. These decisions, which are "unseen, unsigned, and always unexplained," have been used to overturn long-standing precedents. The recent use of the Shadow Docket has undermined the appellate process, given the appearance of partisanship, and produced rigid ideological homogeneity. Professor Vladeck asserts that the Court has used these rulings to advance the Trump administration's policies, while blocking President Biden's policies without establishing precedents that can be used by future democratic presidents.[28]

Professor Vladeck also commented that the decision demonstrates a dubious trend on the part of the Court to grant stays in cases requested by the solicitor general that favor the government's position. Presidents George W. Bush and Barack Obama only sought seven stays from the Supreme Court, while the Trump administration asked for twenty stays and prevailed in most instances. The Supreme Court seems to be encouraging lower courts to accept the federal government's arguments. The lack of standing is often cited as the basis for awarding many of the stays. Professor Vladeck argues that the Court is operating under the "presumption of constitutionality," which assumes that the government is acting in accordance with the law and places the burden of proof on the plaintiff to counter this assumption. The justices appear to be signaling how they would rule on the case when it reaches the Court, rather than addressing the merits of granting the stay. This is causing confusion and undermining the legitimacy of the Court.[29]

In September 2019, the House again voted 236–174 to end the president's emergency declaration. Speaker Pelosi stated: "The administration's decision also dishonors the Congress by negating its most fundamental principle: the separation of powers. It's an assault on our power of the purse."[30] However, the

Senate vote, fifty-four to forty-one, was not sufficient to override the president's veto, so the national emergency continued.

On appeal, the Ninth Circuit in Sierra Club v. Trump held that the Defense Department's (DOD) reprogramming of the $1 billion pursuant to section 8005 into counter drug activities under section 284[31] exceeded its authority and was unconstitutional.[32] The Ninth Circuit in California v. Trump held DOD's reprogramming of $1.5 billion violated the Administrative Procedures Act and was unconstitutional. Notably, the court recognized the danger the wall posed to the Mexican wolf, "including injury, death, harm, and harassment due to construction . . . as well as abandonment of the areas for essential behaviors, such as feeding, resting, and mating due to night lighting and the elimination of food sources and habitat in the area." Furthermore, the wall interfered with cross-border migration, which "is important for increasing and maintaining genetic diversity."[33]

Nevertheless, the Supreme Court refused to lift the stay it placed on the earlier Ninth Circuit decision, which allowed the Trump administration to continue spending these funds on the border wall. Several months later, the Ninth Circuit upheld the permanent injunction issued by the district court in 2019.[34] The Supreme Court agreed to review the Ninth Circuit decision, but canceled the arguments at the request of the Biden administration.

7.2.2 In Re Border Infrastructure Environmental Litigation

The SHS continued to utilize the waiver authority throughout the Trump administration. SHS John Kelley and Acting SHS Elaine Duke invoked their authority under section 102(c) to waive thirty-seven statutes for the construction of 16.2 miles of border wall in the San Diego sector and waived twenty-eight statutes for the construction of 3 miles of replacement fencing in the El Centro sector. The State of California, the DOW, and the CBD brought suits challenging the SHS waivers. The CBD stated: "Trump is willing to throw environmental protections out the window to fulfill his divisive and destructive campaign promise."[35]

In February 2018, the U.S. District Court for Southern District of California decision, In Re: Border Infrastructure Environmental Litigation, held that the SHS waivers of numerous laws, allowing the three projects to proceed, were not ultra vires (outside the scope of their statutory authority). The court also determined that the SHS waiver authority under section 102(c) did not violate the US Constitution.[36] The court was mistaken, repeating the same errors as the earlier court decisions in the Bush administration. The SHS decision was ultra vires (outside the scope of the secretary's authority) because the SHS waiver authority was limited to fencing authorized by the SFA. Because that fencing had been completed in 2013, the SHS waiver authority had ended. In addition, the waiver authority since its enactment violated the nondelegation doctrine.

7.2.2.1 Ultra Vires Step One

The district court acknowledged that there was a strong presumption regarding judicial review of administrative action. This strong presumption could be overcome by "specific language or specific legislative history that is a reliable indicator of congressional intent." The court recognized that Congress precluded judicial review of nonconstitutional violations, but the federal courts still retained authority to determine whether the SHS's waivers were "ultra vires." However, this exception was an "extremely narrow one" and "extraordinary," which the court described as "essentially a Hail Mary pass—and in court as in football, the attempt rarely succeeds." In order to prevail, the plaintiffs must first demonstrate that the secretaries acted in excess of their delegated powers by demonstrating that the issuance of the two waivers contravened the "clear and mandatory" language in section 102. The plaintiffs must also show that precluding judicial review deprived them of any "meaningful and adequate means of vindicating their statutory rights."[37]

The district court, following prior precedents, determined that despite the curtailment of judicial review in section 102(c), it still retained authority to determine whether the SHS waivers were ultra vires.[38] However, the district court failed to recognize that the parameters of the SHS actions were determined not only by the "express language, but also from the structure of the statutory scheme, its objectives, its legislative history and the nature of the administrative action involved."[39] An analysis of the text, intent, and purposes of section 102 provided clear and convincing evidence that the SHS waivers of numerous environmental statutes regarding the replacement fencing and border wall prototypes were ultra vires because they were not authorized under section 102(b).

7.2.2.2 Text

Statutory interpretation begins with the text. The district court did not find the text determinative. The text, however, confined the SHS authority in two ways. The SHS authority was constrained by section 102(b)(1)(A) of the IIRIRA, which was amended by the SFA, requiring the SHS to "construct fencing along not less than 700 miles of the southwest border." This language granted the SHS authority to build at least 700 miles of fencing, the aspirational goal articulated in the SFA. This defined the maximum, not the minimum, amount of authorized fencing.

Section 102 (b)(i)(B) of the IIRIRA instructed the SHS to identify 370 miles of priority areas along the southwest border where fencing had to be constructed to deter smugglers and illegal migrants by December 31, 2008. Secretary Chertoff identified 370 miles of priority areas for border fencing and 300 miles of vehicle barriers, executed the relevant waivers, and committed the DHS to 661 miles of fencing by that date when the SHS authority ended.[40] Once this congressional mandate was met in 2013, the SHS waiver authority should have ended. If the SHS retained unlimited authority to waive any law, at any time, pursuant to section 102(a), the restrictions in (b)(1)(A) and (b)(1)(B) become superfluous.

7.2.2.2.1 Legislative History

Since the district court did not find the text to be clear, the legislative history had to be examined. The district court held that "the parties' varying interpretations concerning the scope of section 102(c) demonstrate the lack of a clear statutory mandate."[41] The legislative history, however, demonstrated that Congress specified the location and amount of fencing in 102(b); consequently, it confined the SHS use of the waiver authority.

The IIRIRA was specifically enacted to complete the construction of 14 miles of border fencing in the San Diego sector. While the bill was being considered, Attorney General Janet Reno stated that the bill was "unnecessary, and we strongly oppose it." President Clinton signed the bill, but registered his skepticism, stating:

> I am, however, extremely concerned about a provision in this bill that could lead to the Federal Government waiving the ESA and NEPA in order to expeditiously construct physical barriers and roads on the U.S. border. I know the Attorney General shares my commitment to those important environmental laws and will make every effort, in consultation with environmental agencies, to implement the immigration law in compliance with those environmental laws.

Secretary of Interior Bruce Babbitt "informed Congress that full compliance with the ESA would not be an impediment to the timely and effective construction of border infrastructure contemplated by this provision." The INS declared that "we will not seek the Attorney General's use of this waiver, and the INS will continue to abide by all environmental laws."[42]

After the completion of the San Diego fencing was delayed, Representative James Sensenbrenner (R. Wisc.), chair of the House Judiciary Committee, inserted the expanded waiver provision in the RIDA amending section 102(c) to allow for the completion of the San Diego fence, which was "still stymied eight years after congressional authorization."[43] Comments by other U.S. Representatives and the Congressional Budget Office's (CBO) budget estimates confirmed that section 102(c) was designed to complete the San Diego fence.[44]

President Bush signed the RIDA into law as an unrelated rider in a supplemental appropriation bill.

Additional border fencing was authorized in the SFA, which amended section 102(b). The SHS was instructed to construct two layers of reinforced fencing along 700 miles in five specific segments along the US–Mexico border. However, doubts were raised regarding the construction of 700 miles of border fencing because of the lack of funding.[45] In addition, President Bush and Senate Republicans were not strong supporters of the aspirational goal of 700 miles of double-layered fencing, but only wanted 370 miles of fencing, 200 miles of vehicle barriers, and 300 miles of electronic surveillance.[46]

Congress loosened the SFA mandate in the Consolidated Appropriations Act of 2008 and scaled back the SHS duties. The double-layered fencing and specific locations for border barriers were eliminated. The SHS was required to construct 700 miles of border fencing, but only if and where it would be "most practical and effective."[47] Supporters of the SFA mandate were disappointed. Representative Peter King (R. N.Y.), the sponsor of the SFA, noted that "the omnibus language guts the Secure Fence Act almost entirely . . . it is unacceptable."[48] There were numerous unsuccessful efforts to restore the SFA mandate to construct 700 miles of double-layered pedestrian fencing, but they all failed.[49]

President Obama in 2011 declared that the southwest border fence is "now basically complete." The DHS reported that 649 of 652 miles described in the SFA had been finished. The vast majority of the SFA-mandated fencing consisted of vehicle barriers and a single-layered pedestrian fence. There were only 36.3 miles of double-layered fence. Upon the completion of the SFA mandate, the SHS waiver authority over congressionally authorized fencing should have ended.

Nevertheless, President Obama noted that many of his Republican opponents would not be satisfied, stating:

> We have gone above and beyond what was requested by the very Republicans who said they supported broader reform as long as we got serious about enforcement. All the stuff they asked for, we've done. But even though we've answered these concerns, I've got to say there are still going to be some who are trying to move the goal posts on us one more time.

President Obama warned: "They'll want a higher fence. Maybe they'll need a moat. Maybe they want alligators in the moat. They'll never be satisfied. And I understand that. That's politics."[50] Attempts to restore double-layered fencing across 700 miles of the southwest border continued in Congress during the Obama administration, but were unsuccessful.[51] These unsuccessful efforts indicated that congressional authorization of additional fencing was necessary to justify the exercise of the SHS waiver authority. Otherwise, these subsequent congressional efforts to restore the SFA mandate and create additional miles of border fence were unnecessary.

7.2.2.2.2 Statutory Purposes

Statutory purposes are an important tool for statutory interpretation. The district court held that the general purpose of section 102(a) of the IIRIRA was to construct border barriers and granted the SHS unlimited authority under section 102(c) to waive any and all laws related to border barrier construction in perpetuity.[52] The court read too much into general statutory purposes and ignored the specific limitations present in the text and legislative history.

Section 102(a), which was entitled a "General" provision, was followed by section 102(b), which provided the DHS with express instructions on the location and the extent of southwest border barriers. Section 102(b) constrained section 102(a) and limited the SHS waiver authority under section 102(c). It is a cardinal rule of "statutory construction that the specific governs the general," especially where "Congress has enacted a comprehensive scheme and has deliberately targeted specific problems with specific solutions." This prevents the specific provision from being "swallowed by the general one." Furthermore, section 102 must be construed so that "no clause, sentence, or word shall be superfluous, void, or insignificant." If section 102(a) granted the SHS unbridled authority to construct border barriers, it would render section 102(b) "insignificant, if not wholly superfluous."[53]

7.2.2.3 *Ultra Vires Step Two*

The district court held that the plaintiffs must also demonstrate that barring judicial review would deprive them of a "meaningful and adequate means to vindicate" their statutory rights to prevail on the ultra vires claim. The district court found that the plaintiffs had not "conduct[ed] a meaningful analysis on this prong." Even if they had, the plaintiffs still failed to meet the requirements of step one to "establish a plain violation of an unambiguous and mandatory provision of section 102."[54] The district court was mistaken. Section 102(a) limited judicial review exclusively to constitutional claims, so the plaintiffs were deprived of any means to question the SHS violation of numerous federal statutes.

7.2.2.4 *Removal of Federal Court Jurisdiction*

Section 102(c) deprived the federal courts of jurisdiction to determine whether the SHS waivers violated numerous statutes. Congress controls federal court jurisdiction over statutory questions. There is a long line of Supreme Court precedents that support congressional power to remove lower federal court jurisdiction over statutory questions. Furthermore, the APA acknowledges that Congress can preclude judicial review of agency action. There is a strong presumption against such removal, but this can be overcome by clear and convincing evidence.[55]

Although Congress has the power to limit the federal courts' jurisdiction, it is not always wise policy. Removing federal court jurisdiction frustrates the constitutional separation of powers, which is designed to prevent the concentration of power in any single branch. In the absence of judicial review, the legislative and executive branches can exercise their power without limit. This undermines judicial authority and frustrates the uniform application of federal law. The public is not able to question federal government action. This denigrates the value of citizen suit provisions in environmental laws, like the ESA, and the idea of the government being limited by law.[56]

Section 102(c) does allow the plaintiffs to raise constitutional issues. Congress cannot deprive federal courts' jurisdiction over constitutional questions. Professor Theodore Eisenberg explained that "it is clear that jurisdictional statutes are subject to constitutional limitations. . . . When their effect is to abrogate constitutional rights, they are no more valid that any other statute violating the Constitution." Congress has the authority to "withdraw jurisdiction from all cases except those in which a particular outcome is mandated by the

Constitution."[57] However, limiting judicial review solely to constitutional issues did not allow plaintiffs to question the SHS adherence to substantive issues protected by environmental statutes.

Judicial review is further hampered because the plaintiffs only have sixty days after the date of the action or the SHS's decision to raise their objections. This is a very short time to assert a constitutional claim, particularly in light of the SHS broad waiver authority. Furthermore, some potential claims might not be ripe within sixty days.

Section 102(c) represents an extreme example of positive political theory, which posits that government institutions behave as rational actors, seeking to have their policy preferences prevail.[58] Congress constrained the federal courts' ability to check the SHS's discretionary authority, clearly intending to prevent any judicial interference in the construction of border barriers.

7.2.2.5 Nondelegation Doctrine

Despite the jurisdictional constraint placed on statutory violations, section 102(c) grants the federal district courts jurisdiction to hear constitutional arguments regarding the SHS waivers. The district court, like earlier courts, did not find that the section 102(c) violated the nondelegation doctrine, which prohibits Congress from delegating its legislative authority to another branch of government.[59]

The nondelegation doctrine is based on Article I, section 1, of the Constitution, which provides "all legislative Powers herein granted shall be vested in a Congress of the United States." The nondelegation doctrine serves to ensure that important social policy is made by Congress, the branch most responsive to the public will. It requires Congress to articulate an intelligible principle to guide the exercise of delegated authority. A statute meets the intelligible principle standard if it "clearly delineates the general policy, the public agency which is to apply it, and the boundaries of the delegated authority."[60]

The district court held section 102(c) set forth a general policy to stop illegal crossings by the construction of roads and barriers. The intelligible principle limiting the delegation of authority was that the SHS may do whatever is necessary for the expeditions construction of border barriers to stop illegal entry.[61] The district court, however, failed to recognize the unprecedented scope of authority delegated in section 102(c) and the lack of any intelligible principles limiting the exercise of such authority. There were no limitations regarding the location of the wall, the scope of the waiver authority, the duration of the waiver, and the geographical area over which the waiver applies. The SHS could waive any and all substantive and procedural rules, including environmental, public health, religious, and other laws, which were outside the SHS's authority and beyond the SHS's expertise.

Two Supreme Court decisions in 1935 struck down statutes for violations of the nondelegation doctrine.[62] Since then the Court has consistently upheld broad delegations of congressional authority.[63] There is an ongoing academic debate regarding the nondelegation doctrine. Supporters argue that responsible administrative decision-making can only be ensured by precise legislative directives.[64] Critics counter that broad delegation promotes accountability and public responsiveness.[65]

Legal scholars, Professors David Barron and Todd Rakoff, noted that the SHS waiver authority under section 102(c) "may be the biggest Congress has yet passed." The statute allocated limitless power to the SHS to waive any and all statutes that interfere with border fence construction. It grants a single agency, the DHS, the ability to waive any and all laws. Furthermore, the SHS is provided with "no standards for picking and choosing among laws to cancel."[66]

The Congressional Research Service recognized the unprecedented scope of discretion granted to the SHS, stating: "After a review of federal law . . . we were unable to locate a waiver provision identical to that of 102 of HR 418 . . . [which] provides a secretary of an executive agency the authority to waive all laws such secretary determines necessary, and directs the secretary to waive such laws."[67]

Congress recognized the unprecedented scope of the section 102(c) waiver. During the House floor debate on the RIDA, Representative Earl Blumenauer (D. Pa.) speculated that the SHS had authority to "give a contract to his political cronies that had no safety standards, using 12-year-old illegal immigrants to do the labor, run it through the site of a Native American burial ground, kill bald eagles in the process, and pollute the drinking water of neighboring communities." He noted that "the amount of unbridled discretion granted is truly extraordinary."[68]

The broad delegation of authority in section 102(c) undermines congressional authority. Public choice scholars view Congress as a marketplace in which public goods are distributed to private interests. Legislation reflects bargains among interest groups that the courts will uphold.[69] The SHS invocation of the waiver undermined the bargains achieved by environmental groups in the congressional arena and deprived them of the ability to challenge these decisions. It also allowed Congress to avoid its constitutional responsibility to make hard policy choices regarding the balance between national security and environmental protection. The "abdication of responsibility is not part of the constitutional design."[70]

Section 102(c) transfers policymaking, a legislative function, to the SHS, an executive official. Executive agencies service and become captured by their clientele. The DHS supports the interests of groups concerned with national security and border barrier construction. Environmental groups are at a strategic disadvantage with the DHS. The broad discretion granted to the SHS regarding the waiver allowed the SHS to provide benefits to the groups supporting border wall construction while undermining the public interest in environmental protection, reinforcing agency capture.[71]

Section 102(c) also violates the principles of federalism. The SHS at her discretion preempted all state, local, and tribal laws that interfered with border wall construction.

Professors Barron and Rakoff note that the unrestrained waiver authority in the RIDA "seems poorly thought out." It granted the SHS broad discretionary power to waive the entire U.S. Code without providing any explanation or procedural protections and precluding judicial review exceeding the limits of the nondelegation doctrine. The RIDA "exemplifies bad statutory design."[72] Congress was solely concerned with building border barriers, while ignoring all other important statutory goals.

7.3 Post-Litigation

7.3.1 Appeal to the Supreme Court

The litigation regarding the SHS waiver authority continued. The RIDA allows a party to appeal constitutional questions to the Supreme Court. Following the district court decision, several environmental groups petitioned for a writ of certiorari. Nine members of the House of Representatives filed an amicus brief, asserting that the DHS violated the Constitution by allowing unelected officials to waive the laws enacted by Congress. However, the Supreme Court denied the petition in December 2018, effectively upholding the district court's decision.[73]

7.3.2 Ninth Circuit Court Appeal

Several appellants renewed their appeal of the Zinke decision in the Ninth Circuit, challenging the SHS's authority pursuant to sections 102 (a) and (b).

The appellants asserted that the district court failed to address the underlying legal predicate of their claim that the SHS's authority to construct border walls under section 102(a) had expired after the goals of section 102(b) were met in 2013. Section 102(a) was limited by section 102(b). In addition, the secretary's decision to proceed with the two projects was arbitrary and capricious because it violated several environmental statutes.[74]

The Ninth Circuit incorrectly upheld the district court decision. The court determined that it retained jurisdiction to consider the appellants' arguments regarding issues that were not based on the secretary's waivers. The Ninth Circuit addressed the "predicate legal question" of the appellants' arguments and found section 102(b) did not limit the SHS authority to construct border barriers under section 102(a). Furthermore, the SHS waivers were not ultra vires because they did not violate "clear and mandatory statutory language."[75]

7.3.3 Center for Biodiversity v. McAleenan

In 2019, several environmental groups brought a similar suit in the U.S. District Court for the District of Columbia, challenging the SHS's waiver of laws allowing for the construction of 20 miles of border wall in New Mexico.[76] The plaintiffs alleged the SHS did not have statutory authority to use its waiver authority because the SHS failed to meet the procedural requirements of section 102(b). Judge Ketanji Brown Jackson rejected this argument on the grounds that judicial authority under section 102(c) was limited to adjudicating constitutional challenges. The court also held the waiver did not violate the nondelegation doctrine,[77] committing the same errors present in the earlier cases.

7.3.4 Final Appeal to the Supreme Court

In their final appeal to the Supreme Court regarding the SHS waivers, several environmental groups filed a petition for a writ of certiorari requesting the Court to review several of the SHS waivers, which allowed the construction of 145 miles of steel bollard walls along the southwest border.[78] The petitioners alleged that the broad waiver authority, coupled with limited judicial review, violated the nondelegation doctrine and the Presentment Clause, because they allowed the partial repeal of existing laws. The CBD declared: "Trump has taken a chainsaw to the Constitution to build wildly divisive border wall. He's done an end run around Congress and waived dozens of laws that protect our air, water, wildlife and public lands. The Constitution prohibits this outrageous executive overreach, and we're asking the Supreme Court to defend these bedrock principles of our democracy."[79] However, the Court denied the petition on June 29, 2020.

7.3.5 Congressional Opposition

There were unsuccessful efforts in Congress to stop President Trump's border wall. Representative Michelle Grisham (D. N.M.) introduced the "Build Bridges Not Walls Act" that prohibited construction of a continuous wall or fence between the United States and Mexico. Representative Adriano Espaillat (D. N.Y.) introduced "This Land is Our Land Act," which prevented the DHS from constructing or expending any funds for the construction of any new border barriers on federal lands within 100 miles of the US border. Representative Kathleen Rice (D. N.Y.) introduced "Rescinding DHS Waiver Authority for Border Wall Act," which terminated the SHS's authority to unilaterally waive "all legal requirements" to construct the border wall.[80]

7.4 The Biden Administration

Border wall construction continued to proceed at a rapid pace in the waning days of the Trump administration.[81] President Biden, on his first day in office, issued Proclamation 10142, which ended President Trump's emergency at the border, halted construction of the border wall for sixty days, and ordered a review of the funding appropriated and redirected for the wall. President Biden withdrew the Trump administration's appeal to the Supreme Court challenging the Ninth Circuit decisions regarding border wall funding. President Biden stated: "It shall be the policy of my administration that no more American taxpayer dollars be diverted to construct a border wall."[82] Nevertheless, some additional border walls were constructed.

7.4.1 Finish the Wall

Funds still remained available for border wall construction. During the Trump administration, Congress had appropriated $1.37 billion in both the FY 2020 and FY 2021 DHS appropriation bills exclusively for the "construction of a barrier system along the southwest border." If these funds were not rescinded by Congress, the DHS would be required to use the funds for their appropriated purposes.

President Biden planned to use the remaining funds to finish parts of the border wall; address safety hazards; install missing components, such as lights, cameras, and detection technology to the incomplete portions of the barrier system; conduct project site restoration; and mitigate environmental and cultural resource damage from border wall construction. President Biden asked Congress to cancel all prior funding for the border wall, but Congress refused to rescind the appropriated funds.[83]

Republicans demanded that the Biden administration spend the appropriated funds to complete the border wall. In March 2021, forty Republican senators asked the General Accounting Office (GAO) to determine whether the Biden administration's failure to spend the funds appropriated for border wall construction violated the Impoundment Control Act (ICA).[84] In June 2021, the GAO determined that the DHS had almost fully obligated the funds appropriated for border wall construction in FY 2018, 2019, 2020. However, most of the FY 2021 appropriations for the border wall remained unobligated due to delays in meeting statutory requirements and developing funding plans. Since the delays were programmatic, President Biden's refusal to spend the appropriated funds did not violate the ICA. The White House budget office declared: "This administration is committed to upholding the rule of law and the President's proclamation directed Federal agencies to comply with appropriations law at every step."[85]

Undeterred, Republicans pursued litigation. The Texas and Missouri attorney generals, along with the Texas Public Land Office, brought suits in the U.S. District Court for the District of South Texas seeking to compel the Biden administration to spend the appropriated funds to resume the building of the border wall.[86] Federal Judge Michaela Alvarez dismissed the suits in August 2022, holding that the parties lacked standing. The parties appealed to the Fifth Circuit, which reversed the district court and remanded the case to the district court for reconsideration.[87]

On remand, Federal Judge Drew Tipton in March 2024 held that the Biden administration's failure to spend the funds appropriated in both FY 2020 and FY 2021 for border wall construction violated the APA. The Biden administration was ordered to spend the appropriations to construct the border wall. The funds could not be diverted to other purposes. Missouri attorney general Andrew

Bailey declared: "The Biden administration has failed to abide by the law to finish the construction of a wall along the southwest border. Joe Biden refuses to carry out his constitutional mandated responsibilities, so we took him to court to force him to do his job."[88]

President Biden, facing an increase in illegal migration and the court order, decided to use some of the appropriated funds to begin the construction of 20 miles of border wall in the Rio Grande Valley. The SHS waived twenty-six laws in southern Texas to permit construction. Environmental groups criticized the decision, arguing the SHS should not have invoked its waiver authority. Instead, the SHS should have been required to comply with environmental laws. The CBD stated: "The Biden administration's decision to allow the wildlife-killing border wall through some of best remaining habitat in Rio Grande Valley is a national disgrace."[89]

The GAO in April 2024 reaffirmed that the Biden administration's failure to spend all of the FY 2020 and FY 2021 appropriated funds for border wall construction did not violate the ICA. The GAO noted that as of January 2024, most of the $1.375 billion appropriated for FY 2020 had been spent. The remaining $12 million balance would remain available until the end of FY 2025. In addition, 47 percent of the $1.375 billion for FY 2021 had not been obligated, but steps were being taken regarding border wall projects. These funds would remain available through FY 2026.[90]

7.4.2 Settlement Agreements

The Biden administration reached settlement agreements in the suits brought by California and the Sierra Club regarding the unlawful diversion of funds to construct the border wall. The administration agreed to provide funds to address the impacts on sensitive habitats and wildlife, including the Mexican wolf, harmed by border wall construction. Passages would be installed in some sections of the border wall and some gates in the wall would remain open to allow wildlife passage. The funding for military construction projects in the plaintiff states would be restored. California's attorney general Rob Bonta declared: "The Trump border wall is officially a relic of the past, which is where it belongs. With environmental mitigation projects coming online to protect our sensitive ecosystem along the U.S.–Mexico border and confirmation of over $427 million in funding restored for military construction projects, today's settlement ushers a new beginning."[91]

7.4.3 Congressional Efforts to Finish the Wall

Republican efforts to complete the wall continued in Congress. In 2023, Representative Bob Good (R. Va.) proposed the "Close Biden's Open Border Act," which diverted $22 billion in UN funding ($11 billion per year) to border wall construction. In 2023, Representative Clay Higgins (R. La.) reintroduced the "Finish the Wall Act," which required the Biden administration to resume border wall construction, prevent any future cancellations of the project, and submit future construction plans to Congress for approval. The 2024 bipartisan Border Act authorized $650 million for border wall construction. None of these bills were enacted.[92]

7.4.4 Mr. Goodbar

The tale of Mr. Goodbar illustrates the problem Mexican wolves have with the border wall. Two Mexican wolves crossed the border in 2017. The female was captured and placed in the captive breeding program. Her pup, Mr. Goodbar, was born in Kansas and released into Arizona in 2020. Mr. Goodbar attempted to cross into Mexico to find a mate, but could not cross because of the border wall. Mr. Goodbar paced along 23 miles of the border wall, and finding no opening, he returned north. The CBD commented: "But beyond one animal's frustrations, the wall separates wolves in the Southwest from those in Mexico and exacerbates inbreeding on both populations." The FWS, diminishing the significance of Mr. Goodbar's failure, noted that "it was speculative to suggest how a barrier may have impacted the wolf's movement." Later, Mr. Goodbar was shot and lost a leg. Despite these challenges, Mr. Goodbar eventually found a mate and is no longer a lone wolf.[93]

Conclusion

Public law litigation regarding the border wall met with both success and failure. Judicial skeptics would applaud the failure of public law litigation in the Bush administration to invalidate the SHS waiver authority, allowing border barrier construction to proceed. The Bush administration utilized the waiver authority five times to allow the construction of 658 miles of border barriers, four of which impacted Mexican wolf migration.

Judicial proponents would support the public law litigation in the Trump administration that halted the unauthorized transfer of funds to build the border wall. Judicial skeptics would support the court decisions upholding the SHS waiver authority, which ended in 2013 and violated the nondelegation doctrine. The SHS utilized the waiver authority twenty-seven times from 2017 to 2021, half of which affected Mexican wolf migration, to authorize the construction of 1,497 miles of border wall not mandated by Congress.[94] However, the Trump administration only built approximately 458 miles of border wall. All of the new barriers were pedestrian barriers, which prevented the passage of large animals, including the Mexican wolf.[95] Most of the construction replaced existing barriers (371 miles), while the remaining construction occurred in new areas (87 miles). Most of the border barrier construction was not fully completed.[96]

The Biden administration promised to stop border wall construction but had to contend with Republican efforts to finish the border wall. The Biden administration settled the litigation regarding President Trump's illegal transfer of funds to build the border wall. The settlements provided funds that would help to further Mexican wolf recovery.

Notes

1 Aaron D. Flesch et al., "Potential Effects of the United States-Mexico Border Fence on Wildlife," 24 Conservation Biology 171, 177–9 (2010).

2 Oscar J. Martinez, "Border Conflict, Border Fences, and the 'Tortilla Curtain' Incident of 1978–1979," 50 Journal of the Southwest 263, 267–8 (2008).

3 Brian P. Segee and Jenny L. Neeley, "On the Line: The Impacts of Immigration Policy on Wildlife and Habitat in the Arizona Borderlands," Defenders of Wildlife (2006), 10. Peter Andreas, "The Escalation of U.S. Immigration Control in the Post-NAFTA Era," 113 Political Science Quarterly 591, 594–8 (Winter 1998–9).

4 Neil F. Devins, "Regulation of Government Agencies Through Limitation Riders," 1987 Duke Law Journal 456, 465 (1987).

5 Pub. L. No. 104–208, Div. C., Title I, section 102 (a), 110 Stat.3009 (1996),

6 Pub. L. No. 109–13, section 102 (c)(1), (2) 119 Stat. 231 (May 11, 2005). Stephen P. Mumme, "The Real ID Act and San Diego-Tijuana Border Fencing," Paper presented at Association of Borderlands Studies Conference, Phoenix, Arizona (April 19–22, 2006), 3, 9–14.

7 P.L.109-367, sect. 3, 120 Stat. 2638 (October 26, 2006).

8 General Accounting Office, *Secure Border Initiative: Technology Deployment Delays Persist and the Impact of Border Fencing Has Not Been Assessed* (September 9, 2009), 8.
9 *Sierra Club v. Ashcroft*, Case No. 04CV0272-LAB (JMA) (S.D. Cal. December 12, 2005), 2.
10 Pub. L. 100–696 (November 18, 1988).
11 527 F. Supp. 2d 119, 123–8 (D.D.C. 2007). Tana M. Sanchez, "Waiving Good-Bye to Environmental Laws Along the Arizona Borderlands," 1 Missouri Environmental Law Review 281, 296–7 (2009).
12 Kate R. Bowers, "Saying What the Law Isn't: Legislative Delegations of Waiver Authority in Environmental Laws," 34 Harvard Law Review 257, 279–82 (2010).
13 533 F. Supp.2d 58, 60–4 (2008).
14 *County of El Paso v. Chertoff*, EP-008-CA-196-FM, 2008 U.S. Dist. LEXIS 83045 (W.D. Texas 2008).
15 Bowers, 279–83.
16 "U.S.–Mexico Fence Building Continues Despite Obama's Promise to Review Effects," Land Letter (April 16, 2009).
17 "Governor Critical of Border Fence Proposal," Associated Press State & Local Wire (December 21, 2005).
18 "Obama Says Border Fence Is 'Now Basically Complete'", PolitiFact (May 16, 2011).
19 "After Legal Challenge FWS Rescind Permit to Trap Cross Border Wolves in Southwest," Silver City (April 11, 2013).
20 Carla N. Argueta, *Border Security: Immigration Enforcement Between Ports of Entry*, Congressional Research Service (April 19, 2016) 19.
21 82 Federal Register 8793 (January 25, 2017).
22 Brief for Plaintiff-Appellees, *El Paso v. Trump*, U.S. Court of Appeals for Fifth Circuit (March 27, 2020), 7–8.
23 Order Granting and Denying Plaintiffs' Motion for Preliminary Injunction, 379 F. Supp. 3d 883, 894–5 (2019).
24 Section 8005 allows the secretary of defense (SOD) to "transfer up to $4 billion of working capital funds of the Department of Defense (DOD) or funds made available in [2019 appropriation act] to the Department of Defense for military functions (except military construction)."
25 *Sierra Club v. Trump*, 379 F. Supp. 3d 883 (N.D. Cal. 2019). *California v. Trump*, 2019 WL 2715421 (N.D. Cal. 2019).
26 *Sierra Club v. Trump*, 929 F.3d 670, 694 (9th Cir. 2019).
27 *Trump v. Sierra Club*, 19A60, 2019 WL 3369425 (U.S. July 26, 2019). "Ruling Allows Shift of Funds to Build Wall," New York Times (July 27, 2019).
28 Stephen Vladeck, *The Shadow Docket* (Basic Books, 2023), 12, 18–21, 157.

29 Stephen Vladeck, "Academic Highlight: The Quiet Doctrinal Shift (likely) Behind Border Wall Stay," Scotus blog (July 27, 2019). See also, Stephen Vladeck, "Power vs. Discretion: Extraordinary Relief and the Supreme Court," Scotus blog (December 12, 2018). Linda Greenhouse, "On the Border Wall, the Supreme Court Caves to Trump," New York Times (August 1, 2019).

30 "House again Votes to Reject Trump's Border Emergency," New York Times (September 28, 2019).

31 Section 284 authorizes the secretary of defense to "provide support for the counterdrug activities . . . of any other department or agency of the Federal government if such support is requested by the official who has responsibility for such counterdrug activities."

32 *Sierra Club, Southern Border Communities Coalition v. Trump*, D.C. Nos. 4: 19-cv-16102, 16300, 00892-HSG (June 26, 2020).

33 *California v. Trump*, D.C. Nos. 19-cv 16299, 00872-HSG (June 26, 2020), 23–4.

34 *Sierra Club v. Trump, California v. Trump*, D.C. Nos. 4: 19-cv-17501, 00892-00172-HSG (October 9, 2020).

35 "Lawsuit Challenges San Diego Border Wall Waiver," Center for Biodiversity (September 6, 2017).

36 284 F. Supp. 3d 1092, 1106–7 (S.D. Cal, 2018). (Curiel)

37 Id., 1110–15.

38 *Leedom v. Kyne*, 358 U.S. 184 (1958). *Spencer Enterprises, Inc. v. United States*, 345 F.3d 683, 689 (9th Cir. 2003); *Amgen v. Smith*, 357 F.3d 103, 113 (D.C. Cir. 2004).

39 *Dart v. U.S.* 848 F.2d 217, 224 (D.C. Cir. 1958).

40 Memorandum in Support of Motion for Summary Judgment By Plaintiffs People of the State of California and the California Coastal Commission (February 9, 2018) 23 (California Memo).

41 Curiel, 1127.

42 Dinah Bear, *Border Wall: Broadest Waiver of Law in American History* (February 2009), 2.

43 151 Congressional Record 1908 (2005).

44 Memorandum of Points and Authorities in support of Plaintiffs Defenders of Wildlife et al. Motion for Summary Judgment (February 9, 2018), 18–19 (DOW Memo). Congressional Budget Office, "Cost Estimate, HR 418-REAL ID Act," 151 Congressional Record H437-02 at H438 (February 9, 2005).

45 "Language in Secure Fences Act Raises Questions about Border Fence," White House Bulletin (October 10, 2006).

46 "Bush Reportedly Wants to Build only Some Portions of Border Fence," Frontrunner (April 3, 2007).

47 P.L. 110-161, div. E, tit. V Sect. 564, 121 Stat. 2090 (December 26, 2007).

48 "Second-chance Defeat When Opponents Couldn't Stop Border-Fence Law, Rendering It Worthless," Augusta Chronicle (December 27, 2007).
49 "Representative Jones Introduces Fence by Date Certain Act," U.S. Fed News (January 17, 2008). "Representative Hunter Introduces Legislation to Reinstate Secure Fences Act," U.S. Fed News (January 23, 2008).
50 Robert Farley, "Obama Says the Border Fence Is 'Now Basically Complete,'" PolitiFact (May 16, 2011).
51 "Senior Deputy Majority Whip Ross Introduces Legislation Directing Completion of Southwest Border Fence," Targeted News (January 15, 2016).
52 Curiel, 116–27.
53 Appellant Opening Brief, 42–3, 48. Rad LAX Gateway Hotel, *LLC v. Amalgamated Bank*, 566 U.S. 639, 645 (2012).
54 Curiel, 1128.
55 *Palmore v. U.S.*, 411 U.S. 389, 400–1 (1973), *Bowen v. Michigan Academy of Family Physicians*, 476 U.S. 667, 670 (1986).
56 Victor M. Sher and Carol Sue Hunting, "Eroding the Landscape, Eroding the Laws: Congressional Exemptions from Judicial Review of Environmental Laws," 15 Harvard Environmental Law Review 482 (1991).
57 Theodore Eisenberg, "Congressional Authority to Restrict Lower Federal Court Jurisdiction," 83 Yale Law Journal 498, 504, 527–58 (1974).
58 William N. Eskridge, "Post-Enactment Legislative Signals," 57 Law and Contemporary Problems 75, 82 (1994); Daniel A. Farber and Phillip P. Frickey, "Forward: Positive-Political Theory in the Nineties," 80 Georgetown Law Journal 457, 462 (1992).
59 Curiel, 1136.
60 *Mistretta v. U.S.*, 488 U.S. 361, 371, 388 (1989).
61 Curiel, 1103–4, 1135–46.
62 *Panama Refining Co. v. Ryan*, 293 U.S. 388 (1935). *A.L.A. Schechter Poultry Co. v. U.S.*, 295 U.S. 495 (1935).
63 Justice Gorsuch noted that the Court continued to police broad delegations under different guises, such as the major questions doctrine, the void for vagueness doctrine, and separation of powers doctrine. *Gundy v. U.S.*, (slip opinion) No 17-6086, 20–2 (June 20, 2019).
64 David Schoenbrod, *Power Without Responsibility: How Congress Abuses the People Through Delegation* (Yale University Press, 1993), 10.
65 Lisa Schultz Bressman, "Schechter Poultry at the Millennium: Delegation Doctrine for the Administrative State," 109 Yale Law Journal 1399, 1406–8 (2000).
66 David J. Barron and Todd D. Rakoff, "In Defense of the Big Waiver," 113 Columbia Law Review 265, 269–72, 290 (2013).

67 Congressional Research Service, *Memo, Section 102 of H.R. 418, Waiver of Laws Necessary for Improvement of Barriers at Borders* (February 9, 2005), 7a–8a.
68 151 Congressional Record H466 (daily ed. February 9, 2005).
69 William N. Eskridge, "Dynamic Statutory Interpretation," 135 University of Pennsylvania Law Review 1479, 1530–1 (1987).
70 *Clinton v. City of New York*, 524 U.S. 417, 450–2 (1998).
71 Id. Bowers, 299–302.
72 Barron and Rakoff, 337–9.
73 Olivia Merritt, "The Wall that Trumps Environmental Law: A Review of Environment and Legal Implications of U.S.–Mexico Border Wall," 32 Villanova Environmental Law Journal 91, 109–10 (2021).
74 *Center for Biodiversity, Defenders of Wildlife, State of California, California Coastal Commission v. U.S. Department. of Homeland Security*, No. 18-55474 (May 14, 2018).
75 U.S. Court of Appeals for the Ninth Circuit, *Center for Biodiversity and Defenders of Wildlife v. U.S. Department of Homeland Security, D.C.* Nos. 3:17-cv-01215, 01873, 01911-GPC-WVG (February 11, 2019).
76 Animal Legal Defense Fund, CBD, DOW, and Southwest Environmental Law Center
77 404 F. Supp. 3d 218 (D.D.C. 2019). Hope Babcock, "Something There Is That Doesn't Love the Wall," 67 Loyola Law Review 13, 28–30 (2021).
78 Center for Biodiversity, Southwest Environment Center, Defenders of Wildlife, Animal Legal Defense Fund.
79 Center for Biodiversity, Petition Asks U.S. Supreme Court to Overturn Border Wall Decisions (January 31, 2020). Babcock, 28–30.
80 "Democrats Float Bills to Block Trump's Wall," Greenwire (February 6, 2017), 17–21. "House Bill Aims to Repeal Waiver Authority for Border Wall," Defenders of Wildlife (March 13, 2019).
81 "Border Wall Construction Forging Ahead," Border Wall Report (November 23, 2020). Bianca Bruno, "Biden and Future of Trump's Border wall," Court House News (November 30, 2020).
82 "Biden Orders a Pause on Border Wall Construction, Bringing Crews to Halt," Washington Post (January 20, 2021). "Supreme Court Cancels Arguments on Trump's Border Wall Policy," New York Times (February 3, 2021).
83 "2020 Appropriations Fall Short of Conservation Needs," Defenders of Wildlife (March 22, 2022).
84 General Accounting Office, "Impoundment Control Act: Use and Impact of Rescission Procedures," testimony of Susan A. Poling (December 16, 2009), 3–4.

85 "Biden's Holdup of Border Funding didn't Violate Law GAO Rules," rollcall.com (June 15, 2021).

86 "Texas Attorney General Ken Paxton sues Biden Administration in bid to Force Construction of U.S.–Mexico Border Wall," Texas Tribune (October 21, 2021). "Missouri Attorney General Files Suit to Require Biden Administration to Build Border Wall," Missouri News (October 21, 2021).

87 "Fifth Circuit Reinstates Red State Lawsuit over Biden's Redirection of Order Wall Funding," Court House News (June 16, 2023).

88 "AG Bailey Obtains Court Order Blocking Biden from Illegally Diverting Funds" (March 13, 2024).

89 "Biden Waives Fed Laws to Allow Border Wall Construction," Associated Press (October 6, 2023).

90 General Accounting Office, *Decision: DHS—Border Barrier Construction and Obligations* (April 22, 2024).

91 Press Release, "AG Bonta Secures Settlement Regarding Unlawful Construction of Trump Border Wall" (July 17, 2023). "Court Settlement Agreement Includes Meaningful Border Wall Changes for Wildlife," AZPM (July 20, 2023).

92 "Freedom Caucus Bill Takes Billions from UN to Finish Border Wall," Fox News (January 11, 2023). Alex Thompson and Hans Nichols, "Harris Flip-Flops on the Border Wall," Axios (August 27, 2024).

93 Martha Pskowski, "Border Wall Thwarts Mr. Goodbar, a Mexican Wolf, from Crossing into Mexico," El Paso Times (December 15, 2021). Martha Pskowski, "Lone Wolf no More," El Paso Times (August 22, 2022).

94 Kenneth D. Madison, *DHS Section 102 (c) Border Waivers* (updated October 5, 2022).

95 General Accounting Office, *Southwest Border Additional Actions Needed to Address Cultural and Natural Resource Impacts from Barrier Construction* (September 7, 2023).

96 General Account Office, *Decision: DHS—Border Barrier Construction and Obligations* (April 22, 2024).

Conclusion

Lobo has faced many travails on his journey from limbo. Different administrations have pursued different policies regarding Mexican wolf recovery. Public law litigation has been instrumental in the return and recovery of the Mexican wolf. Federal courts for the most part have properly interpreted the statutory mandates, both upholding and reversing relevant federal and state actions that have advanced and impeded Mexican wolf recovery. However, public law litigation failed to stop border wall construction, which prevents the cross-border wolf migration necessary for the species' genetic diversity. There were efforts in Congress to halt Mexican wolf recovery and both stop and restart border wall construction, but these efforts were unsuccessful.

Lobo's recovery faced many obstacles. Mexican wolf reintroduction faced opposition from resource developers, particularly livestock producers and hunters. The Mexican wolf was reintroduced into the Blue Range Wolf Recovery Area (BRWRA) as a nonessential experimental population during the Clinton administration. The New Mexico Game and Fish Department (NMGFD) brought suit, challenging the reintroduction. The federal district court in New Mexico Game and Fish Department v. U.S. Fish and Wildlife Service held that the reintroduction complied with the National Environmental Policy Act (NEPA) and Endangered Species Act (ESA). Undeterred, the opponents of wolf reintroduction resurrected their litigation in the George W. Bush administration. The federal district court in Coalition for Arizona/New Mexico Counties for Stable Economic Growth v. U.S. Fish and Wildlife Service dismissed the suit. These decisions allowed Mexican wolf recovery to proceed.

Mexican wolf recovery faced state opposition. The George W. Bush administration established the Adaptive Management Oversight Committee (AMOC), which diminished the U.S. Fish and Wildlife Service's (FWS) authority and elevated state control over wolf management. The AMOC developed policies, particularly State Operating Procedure Thirteen (SOP 13), which adversely affected the wolf population. Environmental groups brought suit challenging the delegation of authority and the AMOC policies. The settlement agreement in Defenders of Wildlife v. Tuggle ended the AMOC and SOP 13, and the FWS regained control over Mexican wolf recovery.

New Mexico pursued policies that jeopardized Mexican wolf recovery during the Obama administration. New Mexico permitted trapping in the BRWRA, which risked harming the small Mexican wolf population. Environmental groups brought suit. The federal district court in Wild Earth Guardians v. Lane upheld the state policy, which violated the ESA and posed a risk to wolf recovery.

New Mexico continued to pursue policies antithetical to Mexican wolf recovery. New Mexico refused to allow the FWS to release wolves into the state until a new recovery plan was put in place. After the FWS refused to comply, the NMGFD brought suit. The federal district court upheld the state's action in New Mexico Game and Fish Department v. U.S. Fish and Wildlife Service, frustrating federal policy. However, the Tenth Circuit reversed the district court and restored the FWS's primary authority over Mexican wolf recovery.

Environmental groups brought suit during the Obama administration seeking to upgrade the status of the Mexican wolf. The settlement agreement in the Defenders of Wildlife v. U.S. Fish and Wildlife Service resulted in the Mexican wolf's status being altered to an endangered subspecies. This necessitated a change in the management regulations.

Mexican wolf recovery presented questions regarding the role of the best available science in federal decision-making mandated by the ESA. Environmental groups brought suit challenging the Obama administration's revised section 10(j) management regulations. The federal district court in Center for Biodiversity v. Jewell halted the regulations because they were not based on the best available science. The Biden administration proposed modified section 10(j) management regulations, which did not address the problems identified by the court in Jewell. Nevertheless, the federal district court in Center for Biodiversity v. Haaland upheld the modified section 10(j) regulations.

The development of Mexican wolf recovery plans, which also involved the use of the best available science, has been contentious. Public law litigation had limited success regarding challenges to the recovery plans. The original minimal recovery plan developed in 1982 was deficient. Scientists developed a scientifically based recovery plan in 2012 during the Obama administration, but it was rejected because of political opposition. The FWS promulgated a revised recovery plan in 2017 during the Trump administration, which was challenged by environmental groups. The federal district court in Center for Biodiversity

v. Zinke upheld parts of the 2017 recovery plan. The court, following previous precedent, held that the recovery plan was discretionary agency action, so was not subject to judicial review. In addition, the court held that the recovery plan did not have to be based on the best available science. However, the court did find that the 2017 recovery plan failed to contain site-specific management actions to adequately address the illegal wolf killings.

Responding to the court's decision, the Biden administration revised the site-specific management actions dealing with the illegal wolf killings. The federal district court in CBD v. Haaland rejected the revised regulations regarding illegal killings, but mistakenly upheld the remainder of the site-specific management actions and the objective measurable criteria in the 2017 recovery plan.

The Biden administration proposed a new recovery plan in 2022. Environmental groups appealed the 2017 Zinke decision in the Ninth Circuit. Environmental groups argued that the new proposed recovery plan was for the most part identical to the prior plan, so the deficiencies in the 2017 plan remained ripe for judicial resolution. The Ninth Circuit in Wild Earth Guardians v. U.S. Fish and Wildlife Service dismissed the appeal, declaring the case was moot. However, many of the issues presented in the 2017 litigation remained unresolved.

Public law litigation was not successful in halting the construction of border barriers or terminating the secretary of homeland security's (SHS) waiver authority. Border barriers interfere with Mexican wolf cross-border migration, which is necessary for the genetic health of the species. Congress instructed the SHS to construct border barriers and waive all laws that interfered with border barrier construction, while severely restricting any meaningful judicial review of the SHS's decisions. The SHS invoked the waiver authority five times in the George W. Bush administration to allow the construction of border barriers. Environmental groups challenged each of the SHS's waivers. Federal district courts consistently upheld the SHS's waiver authority, which violated the nondelegation doctrine. The Supreme Court refused to review the cases.

The Trump administration came to office seeking to expand border wall construction. Congress refused to fund President Trump's border wall. President Trump declared a national emergency and transferred federal funds to construct the border wall. Federal courts determined the transfer of funds violated various statutes. However, the Supreme Court halted the federal court decisions, which allowed unauthorized border wall construction to proceed. Nevertheless, the Ninth Circuit in Sierra Club v. Trump and California v. Trump held that the president's transfer of funds exceeded his statutory authority and was unconstitutional.

Environmental groups continued to challenge the SHS's invocation of the waiver authority. The federal district court decisions In Re Border Infrastructure Environmental Litigation and CBD v. McAleenan continued to uphold the SHS's unfettered waiver authority, allowing border wall construction to proceed in the absence of any environmental protections. The courts failed to acknowledge that the SHS's unlimited waiver authority ended after the border barrier construction authorized by Congress was completed in 2013 during the Obama administration. In addition, the SHS's waiver authority from its inception violated the nondelegation doctrine.

The Biden administration halted border wall construction but had to confront outstanding border wall issues promoted by Republican officials to complete President Trump's border wall.

Judicial skeptics would approve the judicial decisions that upheld government policy. While judicial proponents would support the judicial decisions that carefully analyzed and upheld statutory policies and constitutional restrictions.

8.1 Future Concerns

Lobo is on the road to recovery. The Mexican wolf population grew steadily over the last ten years from 98 in 2015 to 286 in 2024 (162 in New Mexico and 124 in Arizona). In Mexico, the Mexican wolf population increased from thirteen in 2015 to forty-five in 2025. There was also a slight increase in the genetic diversity of both populations.[1] Federal courts played a major role in the return of the Mexican wolf, but many issues remain unresolved.

8.1.1 Cross-Fostering and Adult Releases

Cross-fostering has been beneficial for increasing the Mexican wolf population and addressing genetic problems. Between 2014 and 2023, the FWS cross-fostered ninety-nine pups into forty dens. Only twenty-four of the cross-fostered pups survived their first year. Eighteen cross-fostered pups survived to breeding age, and at least ten have successfully bred, producing more than twenty litters.[2] Nevertheless, the FWS needs to reintroduce more adult bonded pairs with pups. This will provide greater mating opportunities, guaranteeing the wider distribution of valuable genetic material. Wolf families will disperse into suitable habitat, expanding the range of the species. The last release of a family pack occurred in 2006. The FWS reported a 66 percent survival rate for well-bonded pairs with pups released in areas with adequate prey. The FWS has been reluctant to release wolf families because of NMGFD and AZGFD opposition.[3]

8.1.2 Stop Illegal Killings

There has been too much illegal killing of Mexican wolves. Since 1998, 240 Mexican wolves have been killed, 131 illegally. This number does not include those lost to undetected poaching. There were sixty-one suspected illegal takings between 2017 and 2022. Only two individuals were federally prosecuted and three individuals were fined. Human-caused mortality has also harmed wolf recovery in Mexico. From 2011 to 2023, thirty-nine of the forty wolf mortalities were known or likely to have been caused by humans. Many wolves were killed by poisons and vehicle collisions. Greater protection and law enforcement are necessary.[4]

Many Mexican wolves are being killed inadvertently because the shooters believe they are killing a coyote. Juvenile Mexican wolves are similar to coyotes in size, color, and fur patterns, so they are difficult to distinguish from coyotes. Shooters are not prosecuted because the Justice Department follows the McKittrick policy. Shooters will only be prosecuted if they intended to kill a Mexican wolf. Cases of mistaken identity are rarely prosecuted. The McKittrick policy encourages the killing of Mexican wolves. The Department of Justice should stop treating the killing of a Mexican wolf as a specific intent crime and return to its former policy of treating the killing of a Mexican wolf as a general intent crime.[5] This means that for a conviction the government must only prove the shooter knew he was killing an animal, not necessarily a threatened or endangered species for a conviction.

The FWS should not share telemetry frequencies and GPS receivers for Mexican wolf radio collars with members of the public. These devices pinpoint the wolf's location, making it vulnerable to those with malevolent intentions. In 2023, three Mexican wolves were killed by individuals using telemetry devices. These devices should only be provided to federal and state agencies with restrictions on their use.[6]

8.1.3 Protect Coyotes

The killing of coyotes in the Mexican Wolf Experimental Population Area (MWEPA) should be curtailed. Section 4(e) of the ESA allows for the protection of species similar in appearance to threatened or endangered species. Killing coyotes disrupts the pack dynamics and increases the population, resulting in greater livestock depredation. Coyotes kill more livestock than wolves, particularly juvenile lambs and calves. Wolves keep the coyotes in check.[7] Environmental groups asked the FWS to provide ESA protection to coyotes in the MWEPA, which would decrease the killing of Mexican wolves. The FWS refused, asserting that listing the coyote would have a minimum impact on Mexican wolf recovery, impose a heavy burden on law enforcement, and distract from Mexican wolf protection.[8]

8.1.4 Facilitate Cross-Border Migration

Mexican wolves in the United States should have the opportunity to migrate south and join with Mexico's wolf population. The FWS recognized that there are two large blocks of high-quality habitat in Southern Madre Occidental that are connected by areas of lower quality habitat and small isolated patches of high-quality habitat. These areas can support a viable population of Mexican wolves. Connectivity should be established between US and Mexican wolf populations. The FWS concluded that dispersal possibilities were too low to provide adequate gene flow to avoid genetic threats, but acknowledged connectivity was important to maintain the genetic health of the small Mexican wolf populations. Nevertheless, the FWS did recognize that there was a limited possibility for connectivity.[9]

Wolf populations in the MWEPA and northern Sierra Madre Occidental reintroduction sites are approximately 260 miles apart, so they offer dispersal possibilities. High-quality habitat patches in the MWEPA and northern Sierra Madre Occidental are only 96 miles apart. Numerous migratory paths exist between the MWEPA and Sierra Madre Occidental mountain range, particularly through the Sky Islands and the Sonora Desert, which can support a low level of Mexican wolf dispersal.[10] Efforts to facilitate connectivity to enhance genetic diversity should be pursued.

8.1.5 Migration into the Southern Rocky Mountains

Mexican wolves should not be confined to the MWEPA. The FWS must overcome state opposition and allow Mexican wolves to migrate into the Southern Rocky

Mountains (SRM). There are large blocks of public land in northern Arizona, southern Utah, the Grand Canyon, southern Colorado, and northern New Mexico, which provide some of the best wolf habitat in the United States. The 2012 draft recovery plan called for three interconnected populations, each with at least 250 wolves, in the BRWRA, southern Utah/northern Arizona, and southern Colorado/northern New Mexico. Early FWS studies concluded the SRM can support 1,100 wolves. The establishment of a Mexican wolf population in the SRM would create "a spatially segregated population of wolves that extended from the Arctic to Mexico."[11]

Several Mexican wolves, Anubis, Asha, and Ella, have dispersed outside of the MWEPA into northern New Mexico and Arizona. They were either recaptured or shot illegally. These dispersals demonstrate that Mexican wolves must be allowed to migrate beyond I-40 into suitable habitat to promote their recovery.

Colorado recently reintroduced Northern Rocky Mountain (NRM) gray wolves into the state. The FWS should also consider introducing Mexican wolves into southwest Colorado below the NRM wolf release sites. A Mexican wolf population in Colorado would provide a buffer between the NRM and MWEPA populations. Most important, it would provide the opportunity for cross breeding, introducing much needed genetic diversity into the Mexican wolf population.[12]

Recent evidence indicates that the FWS's reliance on a second Mexican wolf population in Mexico to ensure recovery is potentially problematic. Wolves in Mexico are a protected subspecies, but large tracts of federally protected land managed for conservation do not exist in Mexico. The few naturally protected areas that do exist in Sonora, Chihuahua, and Durango are in poor condition and occupied by native populations or rural communities.

Mexico seems unlikely to meet the recovery goal of 100 wolves in Mexico. Since 2007, thirty-five Mexican wolves were released or translocated into Mexico to grow the population and increase its genetic diversity. Nineteen came from the captive population in the United States, and sixteen were translocated from the MWEPA. Unlike in the United States, these releases and translocations occurred in pairs or packs. As of 2023, only nine released or translocated wolves had survived to breeding age.[13]

Only twenty Mexican wolves, six with radio collars, were present at the end of 2023. The lifespan of the wolf in Mexico is on average 214 days, just shy of seven months. However, the median lifespan was only 78 days. Only six of the thirty collared wolves in Mexico lived longer than one year. Only four packs remained in Mexico at the end of 2023. Wolf recovery in Mexico has not met expectations.[14] The FWS should allow dispersal and reintroduction into other regions, particularly the SRM.

8.1.6 Improve Range Policy

The federal government should continue to encourage ranchers to implement adaptive range management practices to prevent the killing of wolves. These measures include altering wolf-livestock grazing rotations, carcass removals, diversionary food caches, hazing, turbo fladry fencing, and range riders. In addition, the federal government should continue to provide compensation to livestock owners for the Mexican wolf's depredation of livestock.[15]

There are several programs that provide compensation for Mexican wolf depredation and financial assistance for proactive, nonlethal activities to address wolf-livestock conflicts. The Wolf-Livestock Loss Demonstration Project Grants (WLLDPG), which is administered by the FWS, provides funds through a competitive process to eligible states and tribes to encourage livestock producers to implement proactive, nonlethal actions to reduce wolf depredation. Funds are also provided to compensate livestock producers experiencing wolf depredation. In addition, the Livestock Indemnity Program, which is administered by the Department of Agriculture's Farm Service Agency, provides compensation to livestock producers for losses in excess of normal mortality due to adverse weather conditions or attacks by reintroduced wolves. These programs have provided compensation for Mexican wolf depredation totaling $950,995 and approved requests for conflict avoidance grants amounting to $734,921.[16]

Compensation plans help to reduce tensions but also provide opportunities for abuse. Since reintroduction, the U.S. Department of Agriculture Wildlife Service (WLS) issued permits for the removal of 206 Mexican wolves. Too many Mexican wolves were removed after being falsely accused of depredation in order to collect compensation, sabotaging recovery efforts. In response, the WLS released a new set of standards to determine the cause of livestock death in Arizona and New Mexico, to prevent Mexican wolves from being unfairly blamed for the killings.[17]

Livestock permittees on public lands should be required to employ nonlethal measures to prevent wolf depredation of livestock. They must be responsible for removing their livestock carcasses, so Mexican wolves do not become habituated to livestock. Mexican wolves feeding on non-wolf killed carcasses should not be removed from the wild. Livestock permittees found guilty of illegal killing or wounding of Mexican wolves should be prosecuted and have their livestock grazing permits revoked.[18]

8.1.7 Stop Border Wall Construction

Border wall construction should not be restarted. Border barriers have caused many problems for Mexican wolf recovery. Border barriers prevent cross-border migration, which is necessary to improve the wolf's genetic diversity. The border barriers in the wildlife corridors that interfere with Mexican wolf migration should be taken down or at least replaced by vehicle barriers, which would permit Mexican wolf migration. Lights along the wall should be removed and noise reduced in areas important for Mexican wolf migration. The DOW and CBD have identified these specific areas.[19] Most important, Congress must rescind the SHS waiver authority and mandate that any border barrier construction, if allowed, must comply with all applicable US environmental laws.

8.1.8 Build the Virtual Wall

The Department of Homeland Security (DHS) should continue its efforts to create a virtual fence along the border, which can distinguish between humans and wildlife. There have been many efforts to create a virtual fence along the southwest border. In November 2005, the DHS launched the Secure Border Initiative Network (SBIN), which was responsible for developing a comprehensive border protection system based on tower-mounted radar and camera technologies.[20] By 2010, the Customs and Border Protection (CBP) at the cost of approximately $1 billion deployed fifteen SBIN tower systems along 53 miles of Arizona's 387-mile border with Mexico.

The Obama administration terminated efforts to create a virtual fence in January 2011. SHS Janet Napolitano stated that "independent, quantitative, science-based review made clear" the virtual fence "cannot meet its original objectives of providing a single, integrated border security technological solution." The 53 miles of virtual fence remained in place, but the funds provided for the high-tech virtual fence were directed to other proven technology.[21]

In the same month, the CBP introduced the Arizona Border Surveillance Technology Plan (ABSTP) for deploying technology along the remainder of the Arizona border. In June 2014, the CBP developed the Southwest Border

Technology Plan that incorporated the ABSTP and extended land-based surveillance technology deployments to the remainder of the southwest border.[22]

The Trump administration continued efforts to create a virtual border fence. Anduril Industries was hired to construct solar-powered mobile surveillance towers with cameras and thermal imaging, which can distinguish humans and wildlife with 97 percent accuracy. "Ghost drones" were developed to enhance its capabilities.[23] Anduril planned to construct 200 towers by 2022. President Trump stated that Anduril's virtual wall system is a "technology so essential it will be a dedicated item in the homeland security budget, reportedly worth several hundred million dollars."[24] The 2021 Omnibus Budget Act allocated $21 million for Anduril surveillance towers and over $200 million more for border security technology to expand autonomous surveillance. In addition, the DHS was authorized to construct a virtual border wall in three locations in south Texas.[25]

President Biden supported the creation of a virtual border wall powered by biometric data, artificial intelligence, facial recognition, aerial drones, infrared cameras, motion sensors, and radar.[26] Since the start of 2022, eighty-three new autonomous surveillance towers have been constructed, bringing the number of surveillance towers to 223. President Biden's FY 2023–4 proposed budget requested $500 million for border technology.[27] The implementation of new technology will enhance the DHS's ability to manage the border while at the same time protecting wildlife.

These measures will further the conservation of the Mexican wolf as an important component of the southwest ecosystem, fulfilling the ESA mandate. However, almost all of these recommendations are unlikely to be implemented in the second Trump administration, which is dedicated to reducing the size of the federal government, cutting federal funds, decreasing protections for endangered and threatened species, stopping immigration, and completing the border wall.

Notes

1 U.S. Fish and Wildlife Service, *2023 5 Year Assessment—2018–2022*. Mexican Wolf Quarterly Update (First Quarter 2025).

2 U.S. Fish and Wildlife Service, "Mexican Wolf Population Grows for Eighth Consecutive Year" (March 5, 2024).

3 Center for Biodiversity et al., Comments: Mexican Wolf Recovery Plan 5-Year Status Evaluation. Letter CBD et al to FWS (June 25, 2024).

4 U.S. Fish and Wildlife Service, "Lower 48-State and Mexico Gray wolf (Canis lupus) Listing, as Revised, 5-Year Review" (2023), 10. U.S. Fish and Wildlife Service, Mexican Wolf Quarterly Update (First Quarter 2025).

5 "Study Suggests Officials Underestimate Wolf Poaching," Santa Fe New Mexican (May 28, 2017). "Endangered Species Policy Flawed: Ruling Ensures Prosecution of Violators," Silver City Sun-News New Mexico (June 23, 2017).

6 "Killing MW Highlights Why Recovery is Failing," Sierra Club (November 20, 2022).

7 Hannah Grover, "Conservation Groups Ask for Fed Protection for Coyotes to Protect Mexican Wolf," New Mexico Political Report (December 13, 2022).

8 "US Gvt Rejects Coyote Protection in Arizona and New Mexico," Axios (April 3, 2023).

9 Mexican Wolf Recovery Plan, Second Revision (2023), 8–9.

10 Earth Justice, "Comments on Draft Recovery Plan," First Revision (2017), 7, 14–26.

11 Michael Robinson, Dave Parsons, and Rob Edward, South from Yellowstone, "What Remains to Be Done," International Wolf (Spring 2006), 8, 9.

12 "Despite Small Increases, Wolves Imperiled by Inbreeding. Isolation," Center for Biodiversity (March 5, 2024). Michael Robinson et al., "Mexican Wolf Recovery Plan 5-Year Status Evaluation," Center for Biodiversity (2022).

13 U.S. Fish and Wildlife Service, 5-Year Evaluation of the Mexican Wolf Recovery Strategy, 16, 22 (December 2024).

14 "After 12 Years of Reintroduction, Population not on Track to Meet Recovery Goals," Center for Biodiversity Letter to U.S. Fish and Wildlife Service (August 22, 2023).

15 86 Federal Register 59953, 59958 (October 29, 2021).

16 U.S. Fish and Wildlife Service, 5-Year Evaluation of the Mexican Wolf Strategy, 13 (December 2024).

17 Spencer Roberts, "Cry Wolf," Intercept (May 24, 2023). "Standards Updated to Confirm Livestock Killings by Wolves in Arizona and New Mexico," Arizona News (September 10, 2023).

18 "Ranchers Say Wolves Threaten their Way of Life," Paysonroundup.com (September 5, 2020).

19 Defenders of Wildlife, *Take Down the Wall: Reconnecting Wildlife Corridors on U.S.–Mexico Border* (2021). Center for Biodiversity, Comments on Border Remediation (June 22, 2023).

20 "Virtual Fence Construction Starts Near Arizona," Associated Press State & Local Wire (May 12, 2009).

21 "Obama Administration Ends High Tech Border Fence," Associated Press Online (January 15, 2011).

22 General Accounting Office, *Southwest Border Security* (November 2017).

23 "Trump Hires Tech Firm to Build a Virtual Border Wall, an Idea Democrats Praised," Washington Post (July 2, 2020). "Anduril Launches a Smarter Drone," TechCrunch (September 10, 2020).

24 "Google AI Tech Will be Used for Virtual Border Wall, CBP Contract Shows," Intercept (October 21, 2020).

25 "Virtual Wall to be Built at 3 Protected Locations in South Texas, Congress Says," Border Report (May 27, 2020).

26 "Biden is Rejecting Trump's Border Wall—but Favors His Own Tech Wall," Truthout (February 2, 2021).

27 Department of Homeland Security, "Fact Sheet: Update on DHS Planning for Southwest Border Security Measures as the 42 Public Health Order Ends" (May 1, 2023).

Table of Cases

A.L.A. Schechter Poultry Co. v. U.S., 295 U.S. 495 (1935).
Alliance for the Wild Rockies v. Salazar, 800 F. Supp. 2d 1123 (2011).
Amgen v. Smith, 357 F.3d 103 (D.C. Cir. 2004).
Boundary Packers v. Boundary County, 913 P.2d 1141 (Idaho, 1996).
Bowen v. Michigan Academy of Family Physicians, 476 U.S. 667 (1986).
California v. Trump, Nos. 19-CV- 00872-HSG, WL 2715421 (N.D. Cal. 2019).
Center for Biodiversity v. Jewell, No. 1;12-CV-1920 (D.D.C. 2013).
Center for Biodiversity et al. v. U.S. Department of Homeland Security, No. 18-55474 (May 14, 2018).
Center for Biodiversity v. Jewell, No. CV-15-00019, No. CV-15-00179, 2018 WL 1586651 (D. AZ., 2018).
Center for Biodiversity v. U.S. Department of Homeland Security, D.C. Nos. 3:17-cv-01215, 01873, 01911-GPC-WVG (9th Cir. February 11, 2019).
Center for Biodiversity v. Haaland, 562 F. Supp. 3d 68 (D. Az. 2021).
Center for Biodiversity v. Haaland, Case 4:22-cv-00303-JAS (July 13, 2022).
Center for Biological Diversity et al. v. Haaland, Case No. 22-15056 (9th Cir. 2023)
Center for Biodiversity v. Haaland, No. CV-22-00303-TUC-SHR (March 31, 2025).
Center for Biodiversity v. McAleenan, 404 F. Supp. 3d 218 (D.D.C. 2019).
Center for Biodiversity v. Zinke, 399 F. Supp. 3d 940 (D. AZ. 2019).
Chemical Manufacturers Assoc. v. Natural Resources Defense Council, 470 U.S. 116 (1985).
Chevron v. Natural Resources Defense Council, 467 U.S. 837 (1984).
Citizens to Preserve Overton Park v. Volpe, 401 U.S. 402 (1971).
Clinton v. City of New York, 524 U.S. 417, (1998).
County of El Paso v. Chertoff, EP-008-CA-196-FM, 2008 U.S. Dist. LEXIS 83045 (W.D. Texas 2008).
Dart v. U.S. 848 F.2d 217 (D.C. Cir. 1958).
Defenders of Wildlife v. Babbitt, 130 F. Supp. 2d 121 (D.D.C. 2001).

Defenders of Wildlife v. Chertoff, 527 F. Supp. 2d 119 (D.D.C. 2007).
Defenders of Wildlife v. Tuggle, 607 F. Supp. 2d 1095 (D. Az., 2009).
Defenders of Wildlife v. Jewell, 2016 U.S. Dist. LEXIS 145122 (D. Az., October 18, 2016).
Defenders of Wildlife v. Zinke, 849 F.3d 1077 (D.C. Cir. 2017).
Duke Power v. Carolina Study Group, 438 U.S. 59 (1978).
Friends of Blackwater v. Salazar, 591 F. 3d 428 (2012).
Fund for Animals v. Babbitt, 903 F. Supp. 96 (D.D.C 1995).
Geer v. Connecticut, 161 U.S. 519 (1896).
Gifford Pinchot Task Force v. U.S. Fish and Wildlife Service, 378 F.3d 1069 (9th Cir. 2004).
Greater Boston Television Corp. v. Federal Communications Commission, 444 F.2d 841 (D.C. Cir. 1970).
Gundy v. U.S., (slip opinion) No 17-6086, (June 20, 2019).
Hughes v. Oklahoma, 441 U.S. 322 (1979).
In Re Infrastructure Environment Border Wall Litigation, 284 F. Supp. 3d 1092 (S.D. Cal, 2018); 915 F.3d 1213 (9th Cir. 2019).
Immigration and Naturalization Service v. Cardoza-Fonseca, 107 S. Ct. 1207, 1224 (1987).
Leedom v. Kyne, 358 U.S. 184 (1958).
Loper Bright Enterprises v. Raimondo, No. 22–451 (June 28, 2024).
Mistretta v. U.S., 488 U.S. 361 (1989).
350 Montana v. Haaland, 50 F.4th 1254 (9th Cir. 2022).
Nevada v. U.S., 512 F. Supp. 166, 172 (1981).
New Mexico Cattle Growers Association v. U.S. Fish and Wildlife Service, Civ. No. 98-367M/JHG (October 28, 1999).
Oregon Natural Resource Council v. Daley, 6 F. Supp. 2d 1139 (D. Or. 1998).
Palmore v. U.S., 411 U.S. 389 (1973),
Panama Refining Co. v. Ryan, 293 U.S. 388 (1935).
Prairie Band of Potawatomi Indians v. Pierce, 253 F.3d 1234 (10th Cir. 2001).
Sierra Club v. Ashcroft, Case No. 04CV0272-LAB (JMA) (S.D. Cal. 2005).
Save Our Heritage v. Chertoff, 533 F. Supp. 2d 58 (2008).
Sierra Club v. U.S. Fish and Wildlife Service, 245 F.3d 434 (5th Cir. 2001).
Sierra Club v. Morton, 405 U.S. 727 (1972).
Sierra Club v. Trump, 379 F. Supp. 3d 883 (N.D. Cal. 2019).
Sierra Club v. Trump, 929 F.3d 670 (9th Cir. 2019).

Sierra Club, Southern Border Communities Coalition v. Trump, D.C. Nos. 4: 19-cv-16102, 16300, 00892-HSG (June 26, 2020).

Sierra Club v. Trump, California v. Trump, D.C. Nos. 4: 19-CV-17501, 00892-00172-HSG (October 9, 2020).

Southwest Center for Biological Diversity v. Bartel, 470 F. Supp. 2d 1118 (S.D. Cal. 2006).

Spencer Enterprises, Inc. v. United States, 345 F.3d 683 (9th Cir. 2003).

Strahan v. Coxe, 127 F.3d 155 (1st Cir. 1997).

Trump v. Sierra Club, 19A60, 2019 WL 3369425 (U.S. July 26, 2019).

TVA v. Hill, 437 U.S. 151 (1973).

U.S. v. McKittrick, 142 F.3d 1170 (9th Circuit 1998).

U.S. v. Students Challenging Regulatory Agency Procedures, 412 U.S. 699 (1973).

Wyoming Farm Bureau v. Babbitt, 987 F. Supp. 1349 (D. Wyo. 1997); 199 F.3d 1224 (10th Cir. 2000).

WildEarth Guardians v. Lane, CIV 12-118 LFG/KBM (December 4, 2012).

WildEarth Guardians et al. v. Haaland, Case No. 22-15029 (9th Cir. 2023).

Bibliography

Alienikoff, T. Alexander, "Updating Statutory Interpretation," 87 *Michigan Law Review* 20 (1988).

Allen, Durward L., *Our Wildlife Legacy* (Funk and Wagnalls, 1962).

Andreas, Peter, "The Escalation of U.S. Immigration Control in the Post-NAFTA Era," 113 *Political Science Quarterly* 591 (Winter 1998–9).

Argueta, Carla N., *Border Security: Immigration Enforcement Between Ports of Entry*, Congressional Research Service (April 19, 2016).

Aubert, Victor, "Competition and Dissensus: Two Forms of Conflict Resolution," 7 *Journal of Conflict Resolution* 26 (1963).

Ausness, Richard C., "Federal Preemption of State Products Liability Doctrines," 44 *South Carolina Law Review* 187 (1993).

Babcock, Hope, "Something There Is That Doesn't Love the Wall," 67 *Loyola Law Review* 13 (2021).

Barron, David J., and Todd D. Rakoff, "In Defense of the Big Waiver," 113 *Columbia Law Review* 265 (2013).

Bauer, Donald C., and William Robert Irvin, eds., *Endangered Species Act: Law, Policy, and Perspectives*, 2d ed. (American Bar Association, 2010).

Bear, Dinah, "Border Wall: Broadest Waiver of Law in American History," Center for International Environmental Law (February 2009).

Bowers, Kate R., "Saying What the Law Isn't: Legislative Delegations of Waiver Authority in Environmental Laws," 34 *Harvard Law Review* 257 (2010).

Bowman, James S., "The Environmental Movement: An Assessment of Ecology Politics," 5 *Environmental Affairs* 649 (1976).

Bressman, Lisa Schultz, "Schechter Poultry at the Millennium: Delegation Doctrine for the Administrative State," 109 *Yale Law Journal* 1399 (2000).

Breyer, Stephen, "Judicial Review of Questions of Law and Policy," 38 *Administrative Law Review* 363 (1986).

Breyer, Stephen, "On the Uses of Legislative History in Interpreting Statutes," 65 *Southern California Law Review* 845 (1992).

Breyer, Stephen, *Reading the Constitution* (Simon & Schuster, 2024).

Brick, Philip D., and R. McGreggor Cawley, eds., *A Wolf in the Garden: The Land Rights Movement and the New Environmental Debate* (Rowman & Littlefield, 1996).

Brown, David E., *The Wolf in the Southwest, The Making of an Endangered Species* (University of Arizona Press, 1984).

Bruskotter, Jeremy, et al., "Expert Judgment and Uncertainty Regarding Protection of Imperiled Species," 31 *Conservation Biology* 657 (June 2017).

Cain, Stanley A., et al., *Predator Control—1971*, Report to the President's Council on Environmental Quality and the Department of Interior by the Advisory Committee on Predator Control (1972).

Callahan, Maureen B., "Must Federal Courts Defer to Agency Interpretations of Statutes? A New Doctrinal Basis for *Chevron U.S.A. v. Natural Resources Defense Council*," 1991 *Wisconsin Law Review* 1275 (1991).

Camacho, Alejandro E., et al., "Assessing State Laws and Resources for Endangered Species Protection," 47 *Environmental Law Reporter* 10837 (2017).

Canon, Bradley, and Charles Johnson, *Judicial Policies: Implementation and Impact*, 2d ed. (Congressional Quarterly, 1998).

Carroll, Carlos, Richard J. Fredrickson, and Robert C. Lacy, "Developing Metapopulation Connectivity Criteria from Genetic Habitat Data to Recover the Endangered Mexican Wolf," 28 *Conservation Biology* 76 (2013).

Carroll, Carlos, et al., "Biological and Sociopolitical Sources of Uncertainty in Population Viability Analysis for Endangered Species Recovery Planning," 9 *Scientific Reports* 10130 (2019).

Chambers, Steven M., et al., "An Account of the Taxonomy of North American Wolves From Morphological Genetic Analysis," 77 *North American Fauna* 1 (October 2012).

Chayes, Abram, "The Role of the Judge in Public Law Litigation," 89 *Harvard Law Review* 1281 (1976).

Chayes, Abram, "The Supreme Court 1981 Term, Public Law Litigation and the Burger Court," 96 *Harvard Law Review* 4 (1982).

Cheever, Federico, "Recovery Planning, the Courts, and the Endangered Species Act," 16 *Natural Resources & Environment* 106 (2001–2).

Cheever, Federico, "The Rhetoric of Delisting Under the Endangered Species Act," 31 *Environmental Law Reporter* 11302 (2001).

Cheever, Federico, "The Road to Recovery, A New Way of Thinking About the Endangered Species Act," 23 *Ecology Law* 1 (1996).

Christensen, Erik, "Genetic Ark: A Proposal to Preserve Genetic Diversity for Future Generations," 40 *Stanford Law Review* 279 (1987).

Coggins, George C., "Federal Wildlife Management," 60 *Oregon Law Review* 59 (1981).

Coggins, George C., "Of Succotash Syndrome and Various Platitudes," 53 *University of Colorado Law Review* 229 (1982).

Coggins, George C., and Parthenia B. Evans, "Predator's Rights and American Wildlife Law," 24 *Arizona Law Review* 821 (1982).

Coppelman, Peter D., "The Federal Government's Response to the County Supremacy Movement," 12 *Natural Resources & Environment* 30 (1997).

Corwin, Wick, "Predator Control and the Federal Government," 51 *North Dakota Law Review* 787 (1975).

Costanza, Robert, et al., "Changes in the Global Value of Ecosystem Services," 26 *Global Environmental Change* 152 (2014).

Costanza, Robert, et al., "The Value of the World's Ecosystem Services and Natural Capital," 387 *Nature* 253 (1987).

Cross, Frank B., "Shattering the Fragile Case for Judicial Review of Rulemaking," 85 *Virginia Law Review* 1243 (1999).

Curtis, Charles, *It's Your Law* (Harvard University Press, 1954).

Czarnezi, Jason J., "An Empirical Investigation of Judicial Decisionmaking, Statutory Interpretations, and the Chevron Doctrine in Environmental Law," 79 *University of Colorado Law Review* 767 (2008).

Davis, Kenneth Culp, *Administrative Law Treatise* (West Law, 1978).

Denvir, John, "Towards a Political Theory of Public Interest Litigation," 54 *North Carolina Law Review* 1133 (1976).

Devins, Neil E., "Appropriations Redux: A Critical Look at Fiscal Year 1988 Continuing Resolution," 1988 *Duke Law Journal* 389 (1988).

Devins, Neil E., "Regulation of Government Agencies Through Limitation Riders," 1987 *Duke Law Journal* 456 (1987).

Dickerson, Reed, *The Interpretation and Application of Statutes* (Little Brown, 1975).

Diver, Colin S., "Statutory Interpretation in the Administrative State," 133 *University of Pennsylvania Law Review* 549 (1985).

Doremus, Holly, "Delisting Engendered Species: An Aspirational Goal, Not a Realistic Expectation," 30 *Environmental Law Reporter* 10434 (2000).

Doremus, Holly, "The Purpose and Effects, and Future of ESA's Best Available Science Mandate," 34 *Environmental Law* 397 (2004).

Doremus, Holly, "Restoring Endangered Species: The Importance of Being Wild," 23 *Harvard Environmental Law Review* 1 (1999).

Dunlap, Thomas R., "Values for Varmints: Predator Control and Environmental Ideas, 1920–1939," 53 *Pacific Historical Review* 141 (1983).

Edvenson, June C., "Predator Control and Regulated Killing: A Biodiversity Analysis," 13 *Journal of Environmental Law* 31 (1995).

Edwards, Lee, ed., *Bringing Justice to the People* (Heritage Books, 2004).

Ehrlich, Paul, and Anne Ehrlich, *Extinction: The Causes and Consequences of the Disappearance of Species* (Random House, 1981).

Eisenberg, Theodore, "Congressional Authority to Restrict Lower Federal Court Jurisdiction," 83 *Yale Law Journal* 498 (1974).

Eskridge Jr., William N., "Dynamic Statutory Interpretation," 135 *University of Pennsylvania Law Review* 1479 (1987).

Eskridge Jr., William N., "The New Textualism," 37 *UCLA Law Review* 621 (1990).

Eskridge Jr., William N., "Overriding Supreme Court Statutory Interpretation Decisions," 101 *Yale Law Journal* 331 (1991).

Eskridge Jr., William N., "Post-Enactment Legislative Signals," 57 *Law and Contemporary Problems* 75 (1994).

Eskridge Jr., William N., and Philip P. Frickey, "Legislation Scholarship and Pedagogy in the Post-Legal Process Era," 48 *University of Pittsburgh Law Review* 691 (1987).

Eskridge Jr., William N., and Philip P. Frickey, "Statutory Interpretation as Practical Reasoning," 42 *Stanford Law Review* 321 (1990).

Farber, Daniel A., and Philip P. Frickey, "Forward: Positive-Political Theory in the Nineties," 80 *Georgetown Law Journal* 457 (1992).

Farina, Cynthia R., "Statutory Interpretation and the Balance of Power in the Administrative State," 89 *Columbia Law Review* 452 (1989).

Fascione, Nina, Aimee Delach, and Martin E. Smith, eds., *People and Predators: From Conflict to Coexistence* (Island Press, 2004).

Fiss, Owen, "The Forms of Justice," 93 *Harvard Law Review* 1 (1979).

Fitzgerald, Edward A., "The Rise and Fall of Worst-Case Analysis," 18 *University of Dayton Law Review* 1 (1992).

Fitzgerald, Edward A., "Wolf Delisting: Old Wine in New Bottles," 44 *Environmental Law Reporter* 10413 (May 2014).

Flesch, Aaron D., et al., "Potential Effects of the United States-Mexico Border Fence on Wildlife," 24 *Conservation Biology* 171 (2010).

Fuller, Lon, "The Forms and Limits of Adjudication," 92 *Harvard Law Review* 92 (1978).

Gambitta, Richard A. L., Marlynn L. May, and James C. Foster, *Governing Through Courts* (Sage, 1981).

Gebbia-Pinetti, Karen M., "Statutory Interpretation: Democratic Legitimacy and Legal System Values," 21 *Seton Hall Legislation Journal* 233 (1997).

Gish, Dan Miles, *An Historical Look at the Mexican Gray Wolf (Canis Lupus Baileyi) in Early Arizona Territory and Since Statehood*, unpublished report, U.S. Fish and Wildlife Services (1977).

Glicksman, Robert L., "Fear and Loathing on the Federal Lands," 45 *University of Kansas Law Review* 647 (1997).

Goble, Dale D., "Evolution of At-Risk Species Protection," in *The Endangered Species Act at Thirty,* Vol. 2, J. Michael Scott, Dale D. Goble, and Frank W. Davis, eds. (Island Press, 2006), 6–23.

Goble, Dale D., "Recovery," in *Endangered Species Act: Law, Policy, and Perspectives,* 2d ed., Donald C. Baur and William Robert Irvin, eds. (American Bar Association, 2010), 70–103.

Greve, Michael S., *The Demise of Environmentalism in American Law* (American Enterprise Institute, 1996).

Grooms, Steve, *The Return of the Wolf* (NorthWord Press, 1993).

Hart, Henry, and Louis Sacks, *The Legal Process: Basic Problems in the Making and Application of Law* (Foundation Press, 1994).

Hedrick, Philip, "Genetics & Recovery Goals for Mexican Wolves," 206 *Biological Conservation* 210 (2017).

Heffelfinger, J. R., et al., "Clarifying Historical Range to Aid Recovery of the Mexican Wolf," 81 *Journal of Wildlife Management* 766 (2017).

Holaday, Bobbie, *Return of the Mexican Wolf* (University of Arizona Press, 2003).

Horowitz, Donald, *The Courts and Social Policy* (Brookings Institution, 1977).

Houck, Oliver A., "The Endangered Species Act and Its Implementation by the U.S. Departments of Interior and Commerce," 64 *University of Colorado Law Review* 277 (1993).

Hungerford, Andrea, "Custom and Culture Ordinances: Not a Wise Move for the Wise Use Movement," 8 *Tulane Environmental Law Journal* 457 (1995).

Jaffe, Louis, *Judicial Control of Administrative Action* (Little Brown, 1965).

Kagan, Robert A., *Adversarial Legalism: The Way of American Law* (Harvard University Press, 2001).

Kagan, Robert A., "Adversarial Legalism and American Government," 10 *Journal of Policy Analysis and Management* 369 (1991).

Keiter, Robert B., and Patrick T. Holscher, "Wolf Recovery Under the ESA: A Study in Comparative Federalism," 11 *Public Land & Resources Law Review* 19 (1990).

Landes, William M., and Richard A. Posner, "The Independent Judiciary in an Interest-Group Perspective," 18 *Journal of Law and Economics* 875 (1975).

Leopold, Aldo, *A Sand County Almanac: And Sketches Here and There* (Oxford University Press, 1966).

Leopold, A. S., et al., *Predator and Rodent Control in the U.S.*, Report submitted to Stewart Udall, Secretary of Interior (March 9, 1964).

Leshy, John D., "Unraveling the Sagebrush Rebellion: Law, Politics and Federal Lands," 14 *University of California Law Review* 317 (1980).

Leventhal, Harold, "Environmental Decisionmaking and the Role of the Courts," 122 *University of Pennsylvania Law Review* 509 (1974).

Lipset, Seymour, ed., *The Third Century: America as a Post-Industrial Society* (Hoover Institution Press, 1979).

Liroff, Richard Allen, "Administrative, Judicial, and Natural Systems: Agency Response to the NEPA of 1969," 3 *Loyola University of Chicago Law Journal* 19 (1972).

Lowi, Theodore, *The End of Liberalism,* 2d ed. (Norton, 1969).

Lund, Thomas A., *American Wildlife Law* (Oxford University Press, 1980).

Macey, Jonathan R., "Promoting Public-Regarding Legislation Through Statutory Interpretation: An Interest-Group Model," 86 *Columbia Law Review* 223 (1986).

Marion, Nancy E., *Making Environmental Law: The Politics of Protecting the Earth* (Praeger, 2011).

Marshall, Thomas R., "Policymaking and the Modern Court: When Do Supreme Court Rulings Prevail?" 42 *Western Political Quarterly* 493 (1989).

Martinez, Oscar J., "Border Conflict, Border Fences, and the 'Tortilla Curtain' Incident of 1978–1979," 50 *Journal of the Southwest* 263 (2008).

McCoy, J. J., *Wild Enemies: The Role of Predators in Nature and in Relation to Man* (Hawthorne Books, 1974).

McCubbins, Matthew D., Roger G. Noll, and Barry R. Weingast, "Positive Canons: The Role of Legislative Bargains in Statutory Interpretation," 80 *Georgetown Law Journal* 705 (1992).

McFarland, Andrew S., *Public Interest Lobbies: Decision Making on Energy* (American Enterprise Institute, 1976).

McNulty, Faith, *Must They Die? The Strange Case of the Prairie Dog and the Black-Footed Ferret* (Doubleday, 1971).

Meese III, Edwin, "Foreword," in *Bringing Justice to the People,* Lee Edwards, ed. (Heritage Books, 2004).

Melnick, R. Shep, "Administrative Law and Bureaucratic Reality," 44 *Administrative Law Review* 245 (1992).

Melnick, R. Shep, *Regulation Through the Courts* (Brookings Institution, 1983).

Merrill, Thomas W., "Judicial Deference to Executive Precedent," 101 *Yale Law Journal* 969 (1992).

Merritt, Olivia, "The Wall that Trumps Environmental Law: A Review of Environment and Legal Implications of U.S.–Mexico Border Wall," 32 *Villanova Environmental Law Journal* 91 (2021).

Michelman, Frank I., "Political Markets and Community Self-Determination: Competing Models of Judicial Legitimacy," 53 *Indiana Law Journal* 145 (1977–8).

Mikva, Abner J., "How Should the Courts Treat Administrative Agencies?" 36 *American University Law Review* 1 (1986).

Miller, Philip S., *Population Viability Analysis for Mexican Wolf: Integrating Wild and Captive Populations in Metapopulation Risk Assessment Model for Recovery Planning,* Report prepared for the U.S. Fish and Wildlife Service (2017).

Molot, Jonathan T., "Judicial Perspectives in the Administrative State," 53 *Stanford Law Review* 1 (2000).

Monaghan, Henry P., "Marbury and the Administrative State," 83 *Columbia Law Review* 1 (1983).

Mortensen, Robert K., *In the Cause of Progress: A History of the New Mexico Cattle Grower's Association* (New Mexico Stockmen, 1983).

Mumme, Stephen P., "The Real ID Act and San Diego-Tijuana Border Fencing," Paper presented at Association of Borderlands Studies Conference, Phoenix, Arizona (April 19–22, 2006).

Newsome, Thomas M., et al., "Top Predators Constrain Mesopredator Distribution," 8 *Nature Communications* 15469 (2017).

Nichol Jr., Gene R., "Justice Scalia, Standing, and Public Law Litigation," 42 *Duke Law Journal* 1141 (1993).

Nie, Martin A., *Beyond Wolves: The Politics of Wolf Recovery and Management* (University of Minnesota Press, 2003).

Nunley, Gary Lee, *The Mexican Gray Wolf in New Mexico* (Division of Animal and Wildlife Control, Department of Interior, 1977).

Paquet, Paul C., John A. Vucetich, Michael K. Phillips, and Leah M. Vucetich, *Mexican Wolf Recovery: Three Year Program Review and Assessment* (June 2001).

Parenteau, Patrick, "Rearranging the Deck Chairs: ESA Reforms in an Era of Mass Extinctions," 22 *William & Mary Environmental Law & Policy Review* 227 (1998).

Parsons, D. R., "Case Study: The Mexican Wolf," 36 *New Mexico Journal of Science* 101 (1996).

Patlis, Jason M., "Recovery, Conservation, and Survival under the ESA: Recovering Species, Conserving Resources, and Saving the Law," 17 *Public Land & Resources Law Review* 55 (1996).

Pedersen Jr., William F., "Formal Records and Informal Rulemaking," 85 *Yale Law Journal* 38 (1975).

Perry, Patrick Austin, "Law West of the Pecos: The Growth of the Wise Use Movement and the Challenge to Federal Public Land-Use Policy," 30 *Loyola Los Angeles Law Review* 275 (1996).

Phillips, Michael K., Rob Edward, and Tina Arapkiles, "Restoring the Gray Wolf to the Southern Rocky Mountains," in *People and Predators: From Conflict to Coexistence*, Nina Fascione, Aimee Delach, and Martin E. Smith, eds. (Island Press, 2004), 240–62.

Pierce, Richard J., "*Lujan v. Defenders of Wildlife*: Standing as a Judicially Imposed Limit on Legislative Power," 42 *Duke Law Journal* 1170 (1993).

Pierce, Richard J., "The Role of the Judiciary in Implementing an Agency Theory of Government," 64 *New York University Law Review* 1239 (1989).

Posner, Richard A., "The Jurisprudence of Skepticism," 86 *Michigan Law Review* 827 (1988).

Primm, Steven A., and Tim W. Clark, "Making Sense of Policy Process for Carnivore Conservation," 10 *Conservation Biology* 1036 (1996).

Prugh, Laura R., et al., "The Rise of the Mesopredator," 59 *Bioscience* 779 (October 2009).

Public Land Law Commission, *One Third of the Nation's Land: A Report to the President and to the Congress* (Washington, DC, 1970).

Rabin, Robert L., "Lawyers for Social Change: Perceptions on Public Interest Law," 28 *Stanford Law Review* 207 (1976).

Rabkin, Jeremy A., *Judicial Compulsions: How Public Law Distorts Public Policy* (Basic Books, 1989).

Radin, Max, "Statutory Interpretation," 43 *Harvard Law Review* 863 (1929–30).

Redish, Martin H., and Theodore T. Chung, "Democratic Theory and the Legislative Process: Mourning the Death of Originalism in Statutory Interpretation," 68 *Tulane Law Review* 803 (1994).

Ripple, William J., and Robert L. Bescheta, "Trophic Cascades in Yellowstone: The First 15 Years After Wolf Reintroduction," 145 *Biological Conservation* 205 (January 2012).

Ripple, William J., et al., "Status and Ecological Effects of World's Largest Carnivores," 343 *Science* 151 (January 2014).

Robinson, Michael, Dave Parsons, and Rob Edward, "South from Yellowstone, What Remains to Be Done," 16 *International Wolf* 8 (Spring 2006).

Robinson, Michael J., *Predatory Bureaucracy: The Extermination of Wolves and the Transformation of the West* (University of Colorado Press, 2005).

Rodgers, William H., *Environmental Law,* 2d ed. (West Publishing Co., 1994).

Rodgers, William H., "Judicial Review of Risk Assessments," 11 *Environmental Law* 301 (1981).

Rodriquez, Daniel B., "The Substance of the New Legal Process," 77 *California Law Review* 919 (1989).

Rohlf, Daniel J., *The Endangered Species Act: A Guide to Its Protections and Implementation* (Stanford Environmental Law Society, 1989).

Rosenbaum, Walter A., *Environmental Policy and Politics*, 3d ed. (Congressional Quarterly, 1995).

Rosenberg, Gerald N., *The Hollow Hope: Can Courts Bring About Social Change?* (University of Chicago Press, 1991).

Ruhl, J. B., "Section 7(a) of the "New" Endangered Species Act," 25 *Environmental Law* 1107 (1995).

Sanchez, Tana M., "Waiving Good-Bye to Environmental Laws Along the Arizona Borderlands," 1 *Missouri Environmental Law Review* 281 (2009).

Santiago-Avila, Francisco J., et al., "Liberalizing the Killing of Endangered Wolves was Associated with More Disappearances of Collared Individuals in Wisconsin, USA," 10 *Scientific Reports* 13881 (2020).

Sax, Joseph L., "The (Unhappy) Truth About NEPA," 26 *Oklahoma Law Review* 239 (1973).

Scalia, Antonin, *A Matter of Interpretation: Federal Courts and the Law* (Princeton University Press, 1997).

Schoenbrod, David, *Power Without Responsibility: How Congress Abuses the People Through Delegation* (Yale University Press, 1993).

Schwartz, Bernard, "Shooting the Piano Player? Justice Scalia and Administrative Law," 47 *Administrative Law Review* 1 (1995).

Scott, J. Michael, Dale D. Goble, and Frank W. Davis, eds., *The Endangered Species Act at Thirty, Vol. 2, Conserving Biodiversity in Human-Dominated Landscapes* (Island Press, 2006).

Segee, Brian P., and Jenny L. Neeley, *On the Line: The Impacts of Immigration Policy on Wildlife and Habitat in the Arizona Borderlands* (Defenders of Wildlife, 2006).

Shapiro, Martin, "Judicial Activism," in *The Third Century: America as a Post Industrial Society*, Seymour Lipset, ed. 109–133 (Hoover Institute, 1979).

Shapiro, Martin, *Who Guards the Guardians? Judicial Control of Administration* (University of Georgia Press, 1988).

Shapiro, Michael E., "Sagebrush and Seaweed Robbery: State Revenue Losses for Onshore and Offshore Federal Lands," 12 *Ecology Law Quarterly* 481 (1985).

Sheingold, Stuart, *The Politics of Rights: Lawyers, Public Policy and Political Change* (Yale University, 1974).

Sher, Victor M., and Carol Sue Hunting, "Eroding the Landscape, Eroding the Laws: Congressional Exemptions from Judicial Review of Environmental Laws," 15 *Harvard Environmental Law Review* 482 (1991).

Smith, Christopher, *Courts, Politics, and the Judicial Process,* 2d ed. (Nelson Hall, 1997).

Soden, Dennis L., *The Environmental Presidency* (SUNY Press, 1999).

Starr, Kenneth W., "Judicial Review in the Post-*Chevron* Era," 3 *Yale Journal on Regulation* 283 (1986).

Starr, Kenneth W., Cass R. Sunstein, Richard K. Willard, and Alan B. Morrison, "Judicial Review of Administrative Action in a Conservative Era," 39 *Administrative Law Review* 353 (1987).

Stewart, Richard, "The Reformation of American Administrative Law," 88 *Harvard Law Review* 167 (1975).

Strauss, Peter L., "Considering Political Alternatives to 'Hard Look' Review," 1989 *Duke Law Journal* 538 (1989).

Sunstein, Cass R., "Constitutionalism After the New Deal," 101 *Harvard Law Review* 421 (1987).

Sunstein, Cass R., "Deregulation and the Hard-Look Doctrine," 1983 *Supreme Court Review* 177 (1983).

Sunstein, Cass R., "Factions, Self-Interest, and the APA: Four Lessons since 1946," 72 *Virginia Law Review* 271 (1986).

Sunstein, Cass R., "Interest Groups in American Public Law," 38 *Stanford Law Review* 29 (1985).

Sunstein, Cass R., "Interpreting Statutes in the Regulatory State," 103 *Harvard Law Review* 405 (1989).

Sunstein, Cass R., "Law and Administration After *Chevron*," 90 Columbia Law Review 2071 (1990).

Sunstein, Cass R., "Naked Preferences and the Constitution," 84 *Columbia Law Review* 1689 (1984).

Sunstein, Cass R., "On the Costs and Benefits of Aggressive Judicial Review of Agency Action," 1989 *Duke Law Journal* 522 (1989).

Sunstein, Cass R., "On the Expressive Function of Law," 144 *University of Pennsylvania Law Review* 2021 (1996).

Sunstein, Cass R., "What's Standing After *Lujan*? Of Citizen Suits, 'Injuries,' and Article III," 91 *Michigan Law Review* 163 (1993).

Titus, A. C., "The Nevada 'Sagebrush Rebellion' Act: A Question of Constitutionality," 23 *Arizona Law Review* 264 (1981).

Todd, David, "Wolf-Predator Control: Endangered Species Protection," 33 *South Texas Law Review* 459 (1992).

Trefethen, James B., *An American Crusade for Wildlife* (Winchester Press, 1975).

Trubek, David M., "Environmental Defense I: Introduction to Interest Group Advocacy in Complex Disputes," in *Public Interest Law: An Economic and Institutional Analysis,* Burton A. Weisbrod, Joel F. Handler, and Neil K. Komesar, eds. (University of California Press, 1978), 151–94.

Urbanski, Mark A., "Chemical Prospecting, Biodiversity Conservation, and the Importance of International Protection of Intellectual Property Rights in Biological Materials," 2 *Buffalo Journal of International Law* 133 (1995).

Vig, Norman J., and Michael E. Kraft, eds., *Environmental Policy in the 1990s* (Congressional Quarterly, 1997).

Vladeck, Stephen, "Academic Highlight: The Quiet Doctrinal Shift (Likely) Behind Border Wall Stay," *Scotus blog* (July 27, 2019).

Vladeck, Stephen, "Power vs. Discretion: Extraordinary Relief and the Supreme Court," *Scotus blog* (December 12, 2018).

Vladeck, Stephen, *The Shadow Docket: How the Supreme Court Uses Stealth Rulings to Amass Power and Undermine the Republic* (Basic Books, 2023).

Wald, Patricia M., "The D.C. Circuit: Here and Now," 55 *George Washington Law Review* 718 (1987).

Wald, Patricia M., "The Sizzling Sleeper: The Use of Legislative History in Construing Statutes in the 1988–89 Term of the US Supreme Court," 39 *American University Law Review* 277 (1990).

Walker, James L., and Michael E. Solomine, "The Next Word: Congressional Response to Supreme Court Statutory Decisions," 65 *Temple Law Review* 425 (1992).

Weisberg, Robert, "The Calabresian Judicial Artist: Statutes and the New Legal Process," 35 *Stanford Law Review* 213 (1983).

Weisbrod, Burton A., "Problems of Enhancing the Public Interest: Toward a Model of Government Failures," in *Public Interest Law*, Burton A. Weisbrod, Joel Handler, and Neil Komesar, eds. (University of California, 1978), 30–41.

Weisbrod, Burton A., Joel F. Handler, and Neil K. Komesar, eds., *Public Interest Law: An Economic and Institutional Analysis* (University of California Press, 1978).

Wellman, Vincent, "Dworkin and the Legal Process Tradition," 29 *Arizona Law Review* 413 (1987).

Wolok, Mimi S., "Experimenting with Experimental Populations," 26 *Environmental Law* 10018 (January 1996).

Young, Stanley P., and Edward A. Goldman, *The Wolves of North America* (Dover, 1944).

Zaleha, D. Bernard, "The Rise and Fall of BLM's Cooperative Management Agreements," 17 *Environmental Law* 125 (1986).

Zellmer, Sandra B., et al., "Species Conservation & Recovery Through Adequate Regulatory Mechanisms," 44 *Harvard Law Review* 367 (2020).

Zeppos, Nicholas S., "Legislative History and the Interpretation of Statutes," 72 *Virginia Law Review* 1295 (1990).

Index

About the Author

Edward A. Fitzgerald is a professor of political science at Wright State University. He has taught at Northeastern University, Saint John's University, the University of Massachusetts Fitchburg, and the University of Dayton Law School. Dr. Fitzgerald received his BA from Holy Cross College in 1971; JD from Boston College Law School in 1974; MA from Northeastern University in 1976, and PhD from Boston University in 1983. He is the author of *The Seaweed Rebellion: Federal-State Conflicts Over Offshore Energy Development* (2001); *Wolves, Courts, and Public Policy: The Children of the Night Return to the Northern Rocky Mountains* (2015), and numerous law review and journal articles.

www.ingramcontent.com/pod-product-compliance
Ingram Content Group UK Ltd.
Pitfield, Milton Keynes, MK11 3LW, UK
UKHW020210290926
15470UKWH00001B/113

9 781666 912821